LTE and LTE Advanced

AF512859

LTE and LTE Advanced

4G Network Radio Interface

André Perez

iSTE

WILEY

First published 2015 in Great Britain and the United States by ISTE Ltd and John Wiley & Sons, Inc.

Apart from any fair dealing for the purposes of research or private study, or criticism or review, as permitted under the Copyright, Designs and Patents Act 1988, this publication may only be reproduced, stored or transmitted, in any form or by any means, with the prior permission in writing of the publishers, or in the case of reprographic reproduction in accordance with the terms and licenses issued by the CLA. Enquiries concerning reproduction outside these terms should be sent to the publishers at the undermentioned address:

ISTE Ltd
27-37 St George's Road
London SW19 4EU
UK

www.iste.co.uk

John Wiley & Sons, Inc.
111 River Street
Hoboken, NJ 07030
USA

www.wiley.com

© ISTE Ltd 2015
The rights of André Perez to be identified as the author of this work have been asserted by him in accordance with the Copyright, Designs and Patents Act 1988.

Library of Congress Control Number: 2015951441

British Library Cataloguing-in-Publication Data
A CIP record for this book is available from the British Library
ISBN 978-1-84821-844-4

Contents

Preface

The aim of this book is to give an integrated presentation of the specifications for Long Term Evolution (LTE) and LTE Advanced radio interfaces, so that the reader can gain an overview of their main characteristics.

The LTE radio interface is the interface between the User Equipment (UE) and the Evolved Packet System (EPS) 4G mobile network.

The LTE radio interface was introduced in Release 8 of the specifications of the 3rd Generation Partnership Project (3GPP).

The EPS mobile network has the following features:

– it provides a data transmission service in Packet-Switched (PS) mode. The data carried by the EPS network are unicast Internet Protocol (IP) packets;

– Release 9 introduced the Multimedia Broadcast Multicast Service (MBMS) using unicast or multicast IP packets;

– Release 9 also introduced the mobile Location Services (LCS).

The LTE radio interface has the following characteristics achieving bitrates of 300 Mbps for the downlink direction and 75 Mbps for the uplink direction:

– 20-MHz radio channel bandwidth;

– 64-Quadrature Amplitude Modulation (64-QAM);

– 4×4 Multiple Input Multiple Output (MIMO) transmission mode for the downlink direction.

The LTE Advanced radio interface was introduced in Release 10.

The LTE Advanced radio interface has the following characteristics achieving bitrates of 3 Gbps for the downlink direction and 1.5 Gbps for the uplink direction:

– 100-MHz radio channel bandwidth, obtained by Carrier Aggregation (CA) of five 20-MHz radio channels;

– 8×8 MIMO transmission mode for the downlink direction;

– 4×4 MIMO transmission mode for the uplink direction.

This book is divided into 10 chapters, summarized in the following table.

Chapters	Title	Contents
1	General characteristics	EPS, MBMS, LCS networks Bearer types Radio interface Network procedures
2	NAS protocol	Attachment Session establishment
3	RRC protocol	System information Connection control Measurements Broadcast control
4	Data link layer	PDCP, RLC and MAC protocols
5	Physical layer	Frequency plan Multiplexing structure Transmission chain
6	Downlink physical signals	PSS, SSS, Cell-Specific RS, MBSFN RS, UE-Specific RS, PRS, CSI RS
7	Downlink physical channels	PBCH, PCFICH, PHICH, PDCCH, PDSCH, PMCH
8	Uplink physical signals	DM-RS, SRS
9	Uplink physical channels	PRACH, PUCCH, PUSCH
10	Radio interface procedures	Access control Data transfer

André PEREZ
September 2015

List of Acronyms

3GPP	3rd Generation Partnership Project
A	
AAA	Authenticate Authorize Answer
AAR	Authenticate Authorize Request
ADM	Activation/Deactivation MAC
AF	Application Function
AKA	Authentication and Key Agreement
AM	Acknowledged Mode
AMBR	Aggregate Maximum Bit Rate
AoA	Angle of Arrival
AP	Application Part
APN	Access Point Name
ARP	Allocation and Retention Priority
ARQ	Automatic Repeat reQuest
AUTN	Authentication Network
B	
BCCH	Broadcast Control Channel
BCH	Broadcast Channel
BI	Backoff Indicator
BM-SC	Broadcast Multicast Service Center
BPSK	Binary Phase Shift Keying
BSR	Buffer Status Report

C

CA	Carrier Aggregation
CC	Component Carrier
CCCH	Common Control Channel
CCE	Control Channel Element
CDD	Cyclic Delay Diversity
CDMA	Code Division Multiple Access
CFI	Control Format Indicator
CGI	Cell Global Identity
CK	Cipher Key
CMAS	Commercial Mobile Alert System
CP	Cyclic Prefix
CPICH	Common Pilot Channel
CRC	Cyclic Redundancy Code
CRI	Contention Resolution Identity
C-RNTI	Cell RNTI
CS	Circuit-Switched
CSFB	Circuit-Switched Fallback
CSI RS	Channel State Information Reference Signal

D

DAC	Digital Analog Converter
DCCH	Dedicated Control Channel
DCI	Downlink Control Information
DL-SCH	Downlink Shared Channel
DM-RS	Demodulation Reference Signal
DNS	Domain Name Server
DRB	Data Radio Bearer
DRX	Discontinuous Reception
DSCP	DiffServ Code Point
DTCH	Dedicated Traffic Channel
DRX	Discontinuous Reception
DwPTS	Downlink Pilot Time Slot

E

EARFCN	E-UTRA Absolute Radio Frequency Channel Number
ECGI	E-UTRAN Cell Global Identifier
E CID	Enhanced Cell Identifier
EMM	EPS Mobility Management
eNB	evolved Node B
EPC	Evolved Packet Core
EPS	Evolved Packet System
E-RAB	EPS Radio Access Bearer
ESM	EPS Session Management
E-SMLC	Evolved Serving Mobile Location Center
ETWS	Earthquake and Tsunami Warning System
E-UTRAN	Evolved Universal Terrestrial Radio Access Network

F

FDD	Frequency Division Duplex
FFT	Fast Fourier Transform
FSTD	Frequency Shift Transmit Diversity

G

GBR	Guaranteed Bit Rate
GERAN	GSM/EDGE Radio Access Network
GMLC	Gateway Mobile Location Center
GNSS	Global Navigation Satellite System
GP	Gap Period
GPRS	General Packet Radio Service
GTPv2-C	GPRS Tunnel Protocol Control
GTP-U	GPRS Tunnel Protocol User
GUMMEI	Globally Unique MME Identity
GUTI	Globally Unique Temporary Identity

H

HARQ	Hybrid Automatic Repeat reQuest
HI	HARQ Indicator
HII	High Interference Indication
HeNB	Home evolved Node B

HRPD	High-Rate Packet Data
HSS	Home Subscriber Server
I	
ICIC	Inter-Cell Interference Coordination
IETF	Internet Engineering Task Force
IFFT	Inverse Fast Fourier Transform
IK	Integrity Key
IMEI	International Mobile Equipment Identity
IMS	IP Multimedia Subsystem
IMSI	International Mobile Subscriber Identity
IOI	Interference Overload Indication
IP	Internet Protocol
L	
LCID	Logical Channel Identifier
LCS	Location Services
LPP	LTE Positioning Protocol
LTE	Long Term Evolution
M	
MAC	Medium Access Control
MAC-I	Message Authentication Code for Integrity
MBMS	Multimedia Broadcast Multicast Service
MBMS GW	MBMS Gateway
MBR	Maximum Bit Rate
MBSFN	MBMS over Single-Frequency Network
MCC	Mobile Country Code
MCCH	Multicast Control Channel
MCE	Multi-cell/Multicast Coordination Entity
MCH	Multicast Channel
MCS	Modulation and Coding Scheme
MIB	Master Information Block
MIMO	Multiple Input Multiple Output
MISO	Multiple Input Single Output
MME	Mobility Management Entity

MNC	Mobile Network Code
MSI	MCH Scheduling Information
MTCH	Multicast Traffic Channel
N	
NAS	Non-Access Stratum
O	
OCC	Orthogonal Covering Code
OFDM	Orthogonal Frequency Division Multiplexing
OFDMA	Orthogonal Frequency Division Multiple Access
OTDOA	Observed Time Difference Of Arrival
P	
PA	Power Amplifier
PBCH	Physical Broadcast Channel
PCCH	Paging Control Channel
PCEF	Policy and Charging Enforcement Function
PCell	Primary Cell
PCFICH	Physical Control Format Indicator Channel
PCH	Paging Channel
PCI	Physical-layer Cell Identity
PCRF	Policy Charging and Rules Function
PDCCH	Physical Downlink Control Channel
PDCCC	Parallel Concatenated Convolutional Code
PDCP	Packet Data Convergence Protocol
PDN	Packet Data Network
PGW	PDN Gateway
PHICH	Physical HARQ Indicator Channel
PHR	Power Head Room
PMCH	Physical Multicast Channel
PMI	Precoding Matrix Indicator
PRACH	Physical Random Access Channel
PRB	Physical Resource Block
P-RNTI	Paging RNTI
PRS	Positioning Reference Signal

PS	Packet-Switched
PSS	Primary Synchronization Signal
PUCCH	Physical Uplink Control Channel
PUSCH	Physical Uplink Shared Channel

Q

QAM	Quadrature Amplitude Modulation
QCI	QoS Class Identifier
QoS	Quality of Service
QPP	Quadratic Permutation Polynomial
QPSK	Quadrature Phase-Shift Keying

R

RAA	Re-Auth Answer
RACH	Random Access Channel
RAPID	Random Access Preample Identifier
RAR	Random Access Response
RAR	Re-Auth Request
RA-RNTI	Random Access RNTI
RAT	Radio Access Technology
RB	Resource Block
RBG	Resource Block Group
RE	Resource Element
REG	Resource Element Group
RFC	Request For Comments
RLC	Radio Link Control
RNTI	Radio Network Temporary Identifier
RNTP	Relative Narrowband Tx Power
ROHC	Robust Header Compression
RRC	Radio Resource Control
RS	Reference Signal
RSCP	Received Signal Code Power
RSRP	Reference Signal Received Power
RSRQ	Reference Signal Received Quality
RSSI	Received Signal Strength Indication

RTT	Round Trip Time
RV	Redundancy Version
S	
SCell	Secondary Cell
SC-FDMA	Single-Carrier Frequency Division Multiple Access
SDF	Service Data Flow
SFBC	Space-Frequency Block Code
SFN	System Frame Number
SGW	Serving Gateway
SIB	System Information Block
SIMO	Single Input Multiple Output
SI-RNTI	System Information RNTI
SISO	Single Input Single Output
SPR	Subscription Profile Repository
SPS	Semi-Persistent Scheduling
SRB	Signaling Radio Bearer
SRS	Sound Reference Signal
SSS	Secondary Synchronization Signal
S-TMSI	Shortened Temporary Mobile Subscriber Identity
T	
TA	Timing Advance
TA	Tracking Area
TAC	Tracking Area Code
TAI	Tracking Area Identity
TC-RNTI	Temporary Cell RNTI
TDD	Time Division Duplex
TEID	Tunnel Endpoint Identifier
TM	Transparent Mode
TMGI	Temporary Mobile Group Identity
TPC	Transmit Power Control
TPMI	Transmitted Precoding Matrix Indicator
TTI	Transmission Time Interval

U

UCI	Uplink Control Information
UE	User Equipment
UICC	Universal Integrated Circuit Card
UL-SCH	Uplink Shared Channel
UM	Unacknowledged Mode
UpPTS	Uplink Pilot Time Slot
USIM	Universal Services Identity Module
UTRAN	Universal Terrestrial Radio Access Network

V

VRB	Virtual Resource Block

1

General Characteristics

1.1. Network architecture

1.1.1. *EPS network*

1.1.1.1. *Functional Architecture*

The Evolved Packet System (EPS) mobile network consists of an Evolved Packet Core (EPC) network and an Evolved Universal Terrestrial Radio Access Network (E-UTRAN) (Figure 1.1).

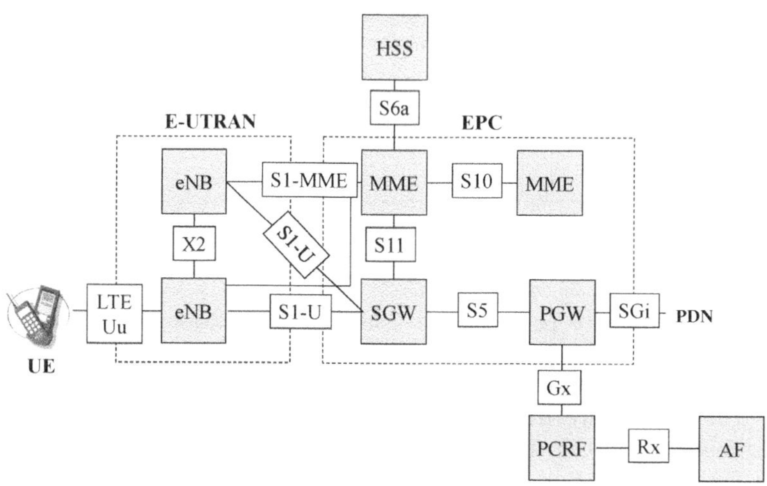

Figure 1.1. *EPS network architecture*

The E-UTRAN access network ensures the connection of the User Equipment (UE).

The EPC core network interconnects the access networks, provides the interface to the Packet Data Network (PDN) and ensures the attachment of the mobiles and the establishment of the bearers.

1.1.1.1.1. eNB entity

The E-UTRAN access network includes a single type of entity, the evolved Node B (eNB) radio station which connects to the mobiles (Figure 1.1).

The eNB entity is responsible for the management of radio resources, for the control of the radio bearer establishment, in which the mobile traffic data is transmitted, and for its mobility management during the handover.

The eNB entity transfers the traffic data from the mobile (respectively from the Serving Gateway (SGW) entity) to the SGW entity (respectively to the mobile).

When the eNB entity receives data from the mobile or the SGW entity, it refers to the QoS Class Identifier (QCI) for the implementation of the data scheduling mechanism.

The eNB entity can perform the marking of the DiffServ Code Point (DSCP) field of IP header, based on the assigned QCI identifier, for outgoing data to the SGW entity.

The eNB entity performs compression and encryption of traffic data on the radio interface.

The eNB entity performs encryption and integrity control of signaling data exchanged with the mobile.

The eNB entity performs the selection of the Mobility Management Entity (MME) to which the mobile is attached.

The eNB entity treats paging requests sent by the MME for their distribution in the cell. The cell is the radio coverage area of the eNB entity.

The eNB entity also distributes system information to the cell containing the technical characteristics of the radio interface, and allowing the mobile to connect.

The eNB entity uses the measurements made by the mobile to decide on the initiation of a cell change during a session (handover).

1.1.1.1.2. MME entity

The Mobility Management Entity (MME) is the network control tower (Figure 1.1). It authorizes mobile access and controls bearer establishment for the transmission of traffic data.

The MME entities belong to a group (pool). Load balancing of MME entities is provided by the eNB entities within a group that must have access to each MME entity of the same group.

The MME entity is responsible for attachment and detachment of the mobile.

During attachment, the MME entity retrieves the subscriber's profile and the subscriber's authentication data stored in the Home Subscriber Server (HSS) entity and performs authentication of the mobile.

During attachment, the MME entity registers the Tracking Area Identity (TAI) of the mobile and allocates a Globally Unique Temporary Identity (GUTI) to the mobile which replaces the private International Mobile Subscriber Identity (IMSI).

The MME entity manages a list of location areas allocated to the mobile, where the mobile can travel in a standby state, without contacting the MME entity to update its TAI location area.

When attaching the mobile, the MME entity selects Serving Gateway (SGW) and PDN Gateway (PGW) entities for the construction of the default bearer, e.g. for access to Internet services.

For the construction of the bearer, the selection of the PGW entity is obtained from the Access Point Name (APN), communicated by the mobile or by the HSS entity in the subscriber's profile.

The source MME entity also selects the target MME entity when the mobile changes both cell and group (pool).

The MME provides the information required for lawful interception, such as the mobile status (standby or connected), the TAI location area if the mobile is in standby or the E-UTRAN Cell Global Identifier (ECGI) of the cell if the mobile is in session.

1.1.1.1.3. SGW entity

The SGW entities are organized into groups (pools). To ensure load balancing of SGW entities, each eNB entity within a group must have access to each SGW entity of the same group.

The SGW entity forwards incoming data from the PGW entity to the eNB entity and outgoing data from the eNB entity to the PGW entity (Figure 1.1).

When the SGW entity receives data from the eNB or PGW entities, it refers to the QCI identifier for the implementation of the data scheduling mechanism.

The SGW entity can perform marking of the DSCP field of IP header based on the assigned QCI identifier for incoming and outgoing data.

The SGW entity is the anchor point for intra-system handover (mobility within the EPS mobile network) provided that the mobile does not change group. Otherwise, the PGW entity performs this function.

The SGW entity is also the anchor point at the inter-system handover PS-PS (Packet-Switched), requiring the transfer of traffic data from the mobile to the 2nd or 3rd generation mobile network.

The SGW entity informs the MME entity of incoming data when the mobile is in standby state, which allows the MME entity to trigger paging of all eNB entities of the TAI location area.

A mobile in the standby state remains attached to the MME entity. However, it is no longer connected to the eNB entity, and thus the radio bearer and the S1 bearer are deactivated.

1.1.1.1.4. PGW entity

The PGW entity is the gateway router providing the EPS mobile network connection to the PDN network (Figure 1.1).

When the PGW entity receives data from the SGW entity or PDN network, it refers to the QCI identifier for the implementation of the data scheduling mechanism.

The PGW entity can perform DSCP marking of IP header based on the assigned QCI identifier.

During attachment, the PGW entity grants an IPv4 or IPv6 address to the mobile.

The PGW entity constitutes the anchor point for inter SGW mobility when the mobile changes groups.

The PGW entity hosts the Policy and Charging Enforcement Function (PCEF) which applies the rules relating to mobile traffic data on packet filtering, on charging and on Quality of Service (QoS) to be applied to the bearer to build.

The Policy Charging and Rules Function (PCRF) entity, outside the EPS mobile network, provides the PCEF function of the PGW entity with the rules to apply when establishing bearers.

The PCRF entity may receive session establishment requests from the Application Function (AF) entity.

The PGW entity generates data for use by charging entities to develop the record tickets processed through the billing system.

The PGW entity performs replication of the mobile traffic data within the framework of lawful interception.

1.1.1.2. *Protocol architecture*

The LTE-Uu interface is the reference point between the mobile and the eNB entity.

This interface supports Radio Resource Control (RRC) signaling exchanged between the mobile and the eNB entity (Figure 1.2) and the mobile traffic data transmitted in the radio bearer (Figure 1.3).

The RRC signaling also provides transport of the Non-Access Stratum (NAS) protocol exchanged between the mobile and the MME entity.

The S1-MME interface is the reference point between the MME and eNB entities for signaling, via the S1-AP (Application Part) protocol.

The S1-AP protocol also provides transport of the NAS protocol exchanged between the mobile and the MME entity (Figure 1.2).

The interface S11 is the reference point between the MME and SGW entities for signaling via the General Packet Radio Service (GPRS) Tunnel Control Protocol (GTPv2-C) (Figure 1.2).

The S5 interface is the reference point between the SGW and PGW entities for signaling via the GTPv2-C protocol (Figure 1.2) and tunneling traffic data (IP packet) via the GPRS Tunnel Protocol User (GTP-U) (Figure 1.3).

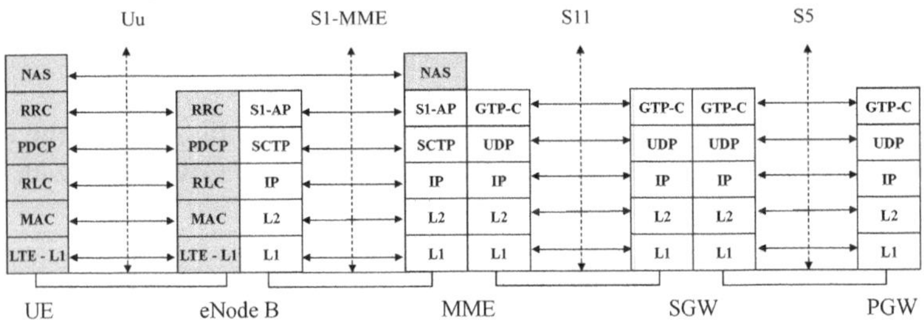

The shaded blocks are subject of a description in the book.
L2 (Layer 2): data link layer
L1 (Layer 1): physical layer

Figure 1.2. *Protocol architecture: control plane*

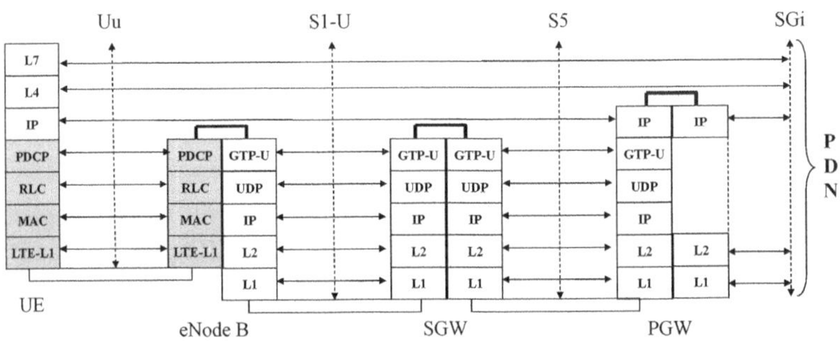

The shaded blocks are subject of a description in the book.
L7 (Layer 7): application layer
L4 (Layer 4): transport layer
L2 (Layer 2): data link layer
L1 (Layer 1): physical layer

Figure 1.3. *Protocol architecture: traffic plane*

The interface S10 is the reference point between the MME entities for signaling, via the GTPv2-C protocol.

The S1-U interface is the reference point between the eNB and SGW entities for tunneling traffic data (IP packet), via the GTP-U protocol (Figure 1.3).

The SGi interface is the reference point between the PDW entity and the PDN data network (Internet) (Figure 1.3).

The X2 interface is the reference point between two eNB entities for signaling, via the X2-AP protocol (Figure 1.4) and tunneling of the mobile traffic data (IP packet), via the GTP-U protocol when mobile changes cells (Figure 1.5).

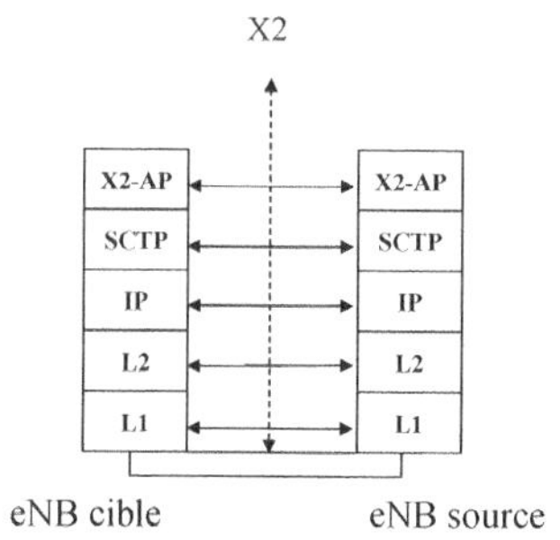

L2 (Layer 2): data link layer
L1 (Layer 1): physical layer

Figure 1.4. *Protocol architecture of the X2 interface: control plane*

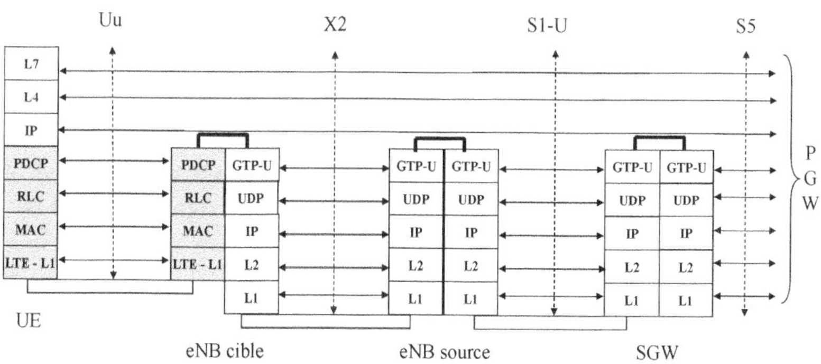

The shaded blocks are the subject of a description in the book.
L7 (Layer 7): application layer
L4 (Layer 4): transport layer
L2 (Layer 2): data link layer
L1 (Layer 1): physical layer

Figure 1.5. *Protocol architecture: traffic plane during handover based on the X2 interface*

The tunnel established between the two eNB entities is unidirectional (eNB source to eNB target). It allows the transfer of traffic data received from the SGW entity to the target eNB entity. It is established temporarily, for the time of the handover of the mobile.

The S6a interface is the reference point between the MME and HSS entities for signaling, via the DIAMETER protocol, enabling access to data from the subscriber (authentication, service profile).

The Gx interface is the reference point between the PCRF and PGW entities for signaling, via the DIAMETER protocol, concerning the transfer of filter rules, the QoS and the charging to be applied to mobile traffic data.

The Rx interface is the reference point between the PCRF and AF entities for signaling via the DIAMETER protocol, concerning session setup requests.

1.1.2. MBMS network

1.1.2.1. *Functional Architecture*

The Multimedia Broadcast Multicast Service (MBMS) network provides a point to multi-point data transport service, for which unicast or multicast IP packets are transmitted from one source to multiple destinations (Figure 1.6).

The MBMS network operates in broadcast mode, and IP packets of the MBMS session are propagated in a multicast bearer independently of mobile requests.

The MBMS over Single Frequency Network (MBSFN) function makes it possible to transmit the same IP packet from multiple synchronized eNB entities. This arrangement improves the quality of the signal received by the mobile.

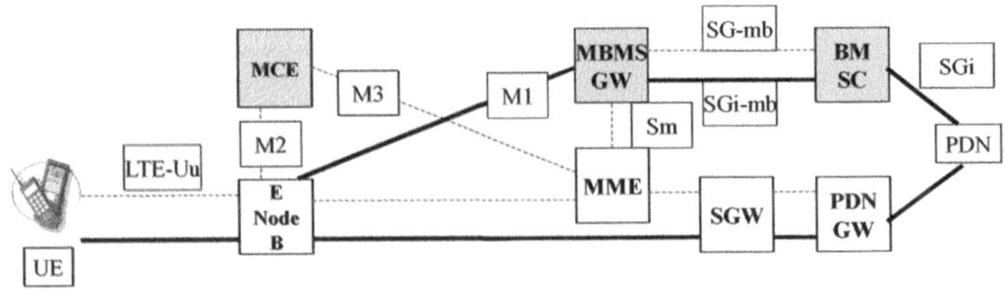

Figure 1.6. *MBMS network architecture*

The MBMS network is composed of different areas:

– the service area (MBMS service area), which determines the set of eNB entities which must transmit the MBMS session;

– the synchronization area (MBSFN synchronization area), which determines a set of synchronized eNB entities. The synchronization area is a subset of the service area;

– the MBSFN area determines a set of coordinated eNB for the simultaneous transmission of a MBMS session. The MBSFN area is a subset of the MBSFN synchronization area. An eNB entity can belong to several MBSFN areas (up to 8);

– the area of reserved cells (MBSFN Area Reserved Cells) determines the eNB entities not involved in the MBSFN transmission sessions.

1.1.2.1.1. BM-SC entity

The Broadcast Multicast Service Centre (BM-SC) entity is the input point of the service stream in the MBMS network.

The BM-SC entity registers the mobile after the authentication procedure.

The BM-SC entity announces the start of the MBMS session to the mobiles.

The BM-SC entity initiates the procedures of starting, modifying and terminating the MBMS session.

The BM-SC entity attributes a Temporary Mobile Group Identity (TMGI) to the session.

The BM-SC entity defines the Quality of Service (QoS) parameters associated with the MBMS session.

The BM-SC entity transmits data using the SYNC protocol that ensures synchronization of their delivery through a set of eNB entities.

1.1.2.1.2. MBMS GW entity

The MBMS Gateway (MBMS GW) entity can be implemented in specific equipment or be integrated with the BM-SC or SGW entities.

The MBMS GW entity allocates an IP multicast address to the bearer for the delivery of data to the eNB entities.

The MBMS GW entity is involved in the procedures of starting, modifying and terminating the MBMS session.

1.1.2.1.3. MCE entity

The Multi-cell/Multicast Coordination Entity (MCE) may be implemented in specific equipment that controls a set of eNB entities or integrated with the eNB entity.

The MCE entity is involved in the procedures of starting, modifying and terminating the MBMS session.

The MCE entity allocates the radio resource to the MBMS session and performs admission control.

The MCE entity defines the Modulation and Coding Scheme (MCS) applied to the radio interface.

The entity MCE performs pre-emption of resources according to the Allocation and Retention Priority (ARP) parameter.

The MCE entity initializes the counting procedure of mobiles involved in the MBMS session.

1.1.2.2. *Protocol architecture*

The SG-mb interface is the reference point between the BM-SC and MBMS GW entities for signaling, via the DIAMETER protocol for starting, modifying or terminating the MBMS session.

The SGi-mb interface is the reference point between the BM-SC and MBMS GW entities for IP packets corresponding to the MBMS session and the SYNC protocol for the eNB entity synchronization.

The Sm interface is the reference point between the MBMS GW and MME entities for signaling via the GTPv2-C protocol for starting, modifying or terminating the MBMS session.

The interface M3 is the reference point between the MME and MCE entities for signaling via the M3-AP protocol for starting, modifying or terminating the MBMS session.

The interface M2 is the reference point between the MCE and eNB entities for signaling via the M2-AP protocol, for the following functions:

– starting, modifying or terminating the MBMS session;

– counting of terminals subscribed to an MBMS session;

– configuring the physical channel on the radio interface.

The M1 interface is the reference point between, on the one hand, the MBMS GW entity, and, on the other hand, all eNB entities involved in the distribution of the MBMS session for tunneling traffic data (the IP packet of the MBMS session), via the GTP-U protocol.

1.1.3. *LCS network*

1.1.3.1. *Functional Architecture*

The Location Services (LCS) network provides the location service of the mobile, and optionally the speed of the movement (Figure 1.7).

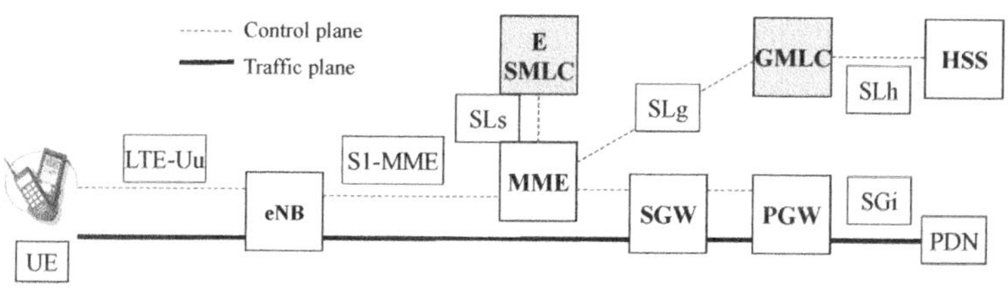

Figure 1.7. *LCS network architecture*

The LCS network uses different methods to locate the mobile:

– the Observed Time Difference Of Arrival (OTDOA) mechanism consists of measuring the time difference at the mobile between the reception of two different signals;

– the Global Navigation Satellite System (GNSS) based on satellites;

– the Enhanced Cell Identifier (E-CID) mechanism based on measurements made by the eNB entity and the mobile, such as Reference Signal Received Power (RSRP), Round Trip Time (RTT) or Angle of Arrival (AoA).

1.1.3.1.1. E-SMLC entity

The Evolved Serving Mobile Location Centre (E-SMLC) entity implements the location of the mobile.

The E-SMLC entity coordinates the use of resources and dialogue with the mobile and the eNB entity to retrieve the information used to locate the mobile.

In the mode based on the mobile (UE-based), the mobile performs the measurements and the estimation of its location.

In the mode where the mobile is assisted (UE-assisted), the mobile performs measurements and the E-SMLC entity provides the estimated location of the mobile.

In the mode in which the eNB entity is assisted (eNB-assisted), the eNB entity performs measurements and the E-SMLC entity provides the estimated location of the mobile.

Table 1.1 provides the correspondence between the methods and the mobile positioning modes.

Methods	UE-based	UE-assisted	eNB-assisted
OTDOA	no	yes	no
GNSS	yes	yes	no
E-CID	no	yes	yes

Table 1.1. *Methods and mobile positioning modes*

1.1.3.1.2. GMLC entity

The Gateway Mobile Location Centre (GMLC) entity is the LCS network access point of an external client wishing to obtain the location information of a mobile.

The GMLC entity recovers the identity of the MME entity which attaches the mobile from the HSS entity, then transmits the mobile location request and retrieves the mobile location information.

1.1.3.2. *Protocol architecture*

The SLs interface is the reference point between the E-SMLC and MME entities for signaling, via the LCS-AP protocol, relating to the mobile location request.

The LTE Positioning Protocol (LPP) exchanged between the mobile and the E-SMLC entity is carried by the LCS-AP and NAS protocols (Figure 1.8).

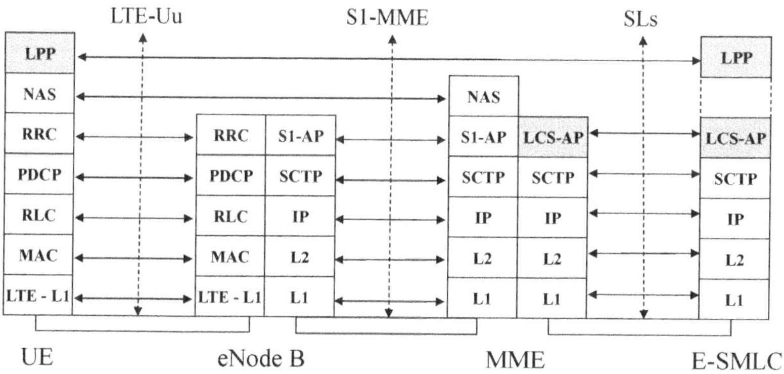

Figure 1.8. *Protocol architecture: LLP protocol transport*

The LPPa protocol exchanged between the eNB and E-SMLC entities is carried by the LCS-AP and S1-AP protocols (Figure 1.9).

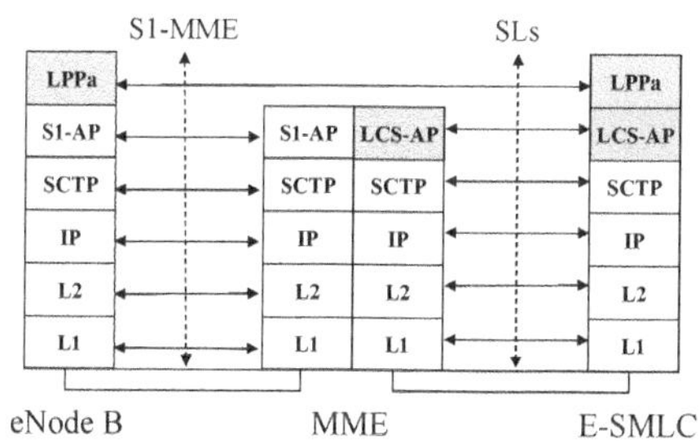

Figure 1.9. *Protocol architecture: LLPa protocol transport*

The LPP and LPPa protocols allow the exchange of information relating to the mobile location mechanism.

The SLg interface is the reference point between the MME and the GMLC entities for signaling, via the DIAMETER protocol, relating to the mobile location request.

The SLh interface is the reference point between the GMLC and HSS entities for signaling, via the DIAMETER protocol, allowing recovery of the identity of the MME entity that attached the mobile.

1.2. Bearer types

1.2.1. *Bearer structure*

The EPS mobile network carries the traffic data (IP packet) transparently to the entity PDN Gateway (PGW) that performs packet routing.

The Internet Protocol (IP) packet is carried in bearers constructed between EPS mobile network entities (Figure 1.10).

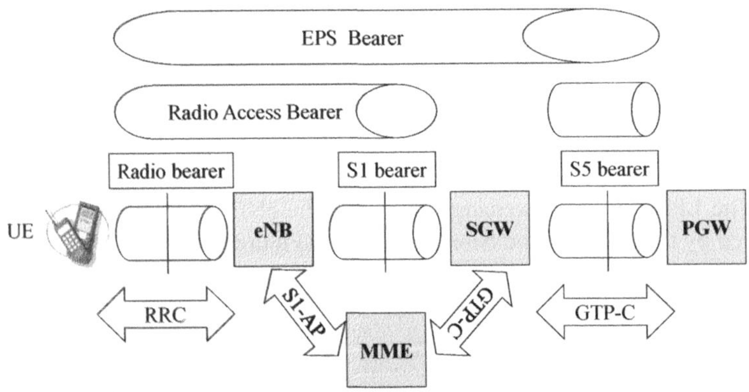

Figure 1.10. *Construction of the bearers*

The Data Radio Bearer (DRB) is constructed between the User Equipment (UE) and the eNB entity. The Radio Resource Control (RRC) signaling, which is exchanged between the mobile and the eNB entity, is responsible for constructing this bearer.

The S1 bearer is constructed between the eNB and SGW entities. The S1-AP signaling, exchanged between the eNB and MME and the GTPv2-C signaling, exchanged between the MME and SGW entities, are responsible for constructing this bearer.

The S5 bearer is constructed between the SGW and PGW entities. The GTPv2-C signaling exchanged between the SGW and PGW entities is responsible for constructing this bearer.

The connection of the radio bearer and the S1 bearer, performed by the eNB entity, constitutes the EPS Radio Access Bearer (E-RAB).

The connection of the E-RAB and S5 bearers, which is performed by the SGW entity, constitutes the EPS bearer.

The PGW entity is the only EPS mobile network entity that routes the mobile traffic data (IP packets).

The IP transport network that enables communication between the EPS network entities routes the S1 or S5 bearers.

The eNB and SGW entities do not perform routing. They only provide the connection between the bearers.

There are two types of bearers in the EPS mobile network:

– the default bearer established when attaching the mobile;

– the dedicated bearer established following a specific request from the mobile.

The traffic data is transmitted in an EPS bearer. A bearer may carry multiple traffic data with the same QoS.

1.2.2. *Quality of Service*

The EPS bearer can be of Guaranteed Bit Rate (GBR) type or can be of non-GBR type.

Table 1.2 provides the QoS characteristics associated with these two bearer families.

QoS characteristics	GBR	Non-GBR
QCI (QoS Class Identifier)	✓	✓
ARP (Allocation and Retention Priority)	✓	✓
GBR (Guaranteed Bit Rate)	✓	
MBR (Maximum Bit Rate)	✓	
APN-AMBR (Aggregate Maximum Bit Rate)		✓
UE-AMBR		✓

Table 1.2. *QoS characteristics*

The QCI parameter indicates the priority level, the delay and the packet loss rate (Table 1.3).

QCI from 1 to 4 are assigned to GBR bearers.

QCI from 5 to 9 are assigned to non-GBR bearers.

QCI	Resource Type	Priority	Delay	Packet loss rate	Examples of services
1.	GBR	2	100 ms	10^{-2}	Voice
2.		4	150 ms	10^{-3}	Video Calling
3.		3	50 ms	10^{-3}	Games
4.		5	300 ms	10^{-6}	Video.
5.	Non-GBR	1	100 ms	10^{-6}	SIP Signaling
6		6	300 ms	10^{-6}	Video, Internet
7		7	100 ms	10^{-3}	Voice, Video, Games
8		8	300 ms	10^{-6}	Video, Internet
9		9			

Table 1.3. *QCI parameter characteristics*

The scheduling of traffic data carried out at the level of the eNB, SGW and PGW entities is based on the QCI priority level.

The bit rate control is done from the GBR and MBR parameters for guaranteed bit rate bearers.

The control is conducted for each bearer at the eNB and PGW entities for incoming data in the EPS mobile network.

The bit rate control is done from the APN-AMBR and UE-AMBR parameters for non-GBR type bearers. This control is performed for aggregated bit rates of non-GBR bearers of a mobile.

The APN-AMBR parameter controlled by the PGW entity corresponds to the maximum bit rate authorized for all the streams of mobile using non-GBR bearers, at the PGW entity level.

The UE-AMBR parameter controlled by the eNB entity corresponds to the maximum authorized bit rate for all streams of mobile using non-GBR bearers, at the eNB entity level.

The pre-emption implemented at the eNB and PGW entity level corresponds to the ARP parameter that defines the following information:

– pre-emption capability: this parameter is used for the establishment of a new session, if the resource is not available. This parameter determines whether or not a new session can pre-empt an existing session;

– pre-emption vulnerability: this parameter is used by the existing session. This parameter is compared to the Pre-emption Capability parameter of the new session to determine whether the existing session can be pre-empted or not;

– priority: this determines the priority level associated with pre-emption. This priority level is independent of that set for the QCI parameter.

The QoS parameters (QCI, ARP and APN-AMBR) relating to default bearers are stored in the Home Subscriber Server (HSS) entity. These values can be changed by the Policy and Charging Rules Function (PCRF) entity.

The QoS parameters (QCI, ARP, GBR and MBR) relating to dedicated bearers are stored in the Subscription Profile Repository (SPR) entity associated with the PCRF entity.

The MME entity replaces the UE-AMBR parameter provided by the HSS entity by the sum of the different APN-AMBR parameters, provided it is less than the value indicated by the HSS entity.

1.3. Radio interface

The LTE radio interface was introduced in Release 8 of the specifications of the 3rd Generation Partnership Project (3GPP).

The LTE Advanced radio interface was introduced in Release 10.

1.3.1. *Structure of the radio interface*

At the LTE-Uu radio interface, between the User Equipment (UE) and the eNB entity, the traffic data, corresponding to an IP packet, and, the signaling data, corresponding to a Radio Resource Control (RRC) message, are encapsulated by the data link layer broken down into three sublayers (Figure 1.11):

– the Packet Data Convergence Protocol (PDCP);

– the Radio Link Control (RLC) protocol;

– the Medium Access Control (MAC) protocol.

Three types of channels are defined (Figure 1.11):

– the logical channel defines the data structure at the interface between the RLC and MAC sublayers;

– the transport channel defines the data structure at the interface between the MAC sublayer and the physical layer;

– the physical channel defines the data structure between the two parts making up the physical layer, first, the coding, and secondly the modulation and the multiplexing.

The RRC messages can carry the Non-Access Stratum (NAS) messages exchanged between the mobile and the MME entity.

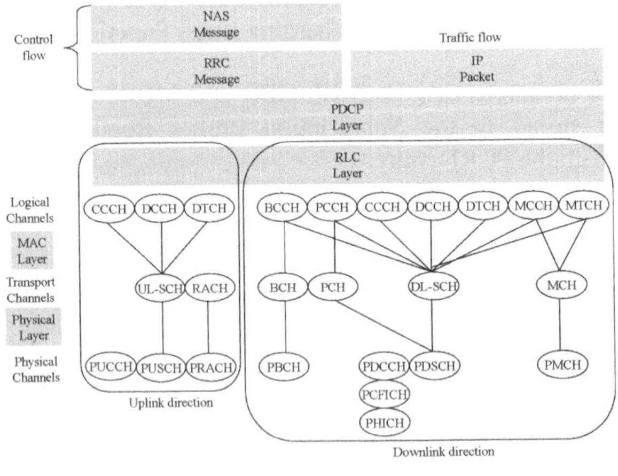

Figure 1.11. *Radio interface structure*

1.3.2. *NAS protocol*

The NAS protocol is the signaling exchanged between the mobile and the MME entity. It is transported by the RRC protocol over the LTE-Uu radio interface and by the S1-AP protocol over the S1-MME interface.

The NAS protocol consists of two protocols:

– the EPS Mobility Management (EMM) protocol is responsible for the control of mobility and security;

– the EPS Session Management (ESM) protocol is responsible for the control of session establishment.

The mobile can be in two operating states, either the registered state (EMM-REGISTERED), or the deregistered state (EMM-DEREGISTERED).

In the deregistered state, the mobile is not attached to the MME, and therefore it cannot be contacted. The move to the registered state is done by the attachment of the mobile which comprises the following four procedures:

– mutual authentication of the mobile and the MME entity;

– registration of the Tracking Area Identifier (TAI) of the mobile to the MME entity;

– provision of the Globally Unique Temporary Identifier (GUTI) to the mobile;

– establishment of the default bearer.

The transition to the deregistered state occurs if the mobile is detached or if the attachment of the mobile or the updating of the location or the service request is rejected by the MME entity.

In addition to the EMM-REGISTERED state, the mobile can be in two operational states, the idle state (ECM-IDLE) or the connected state (ECM-CONNECTED).

In the idle state, the transport of NAS protocol, the RRC protocol on the LTE-Uu interface or the S1-AP protocol on the S1-MME interface are not established.

In the connected state, the transport of NAS protocol, the RRC protocol on the LTE-Uu interface and the S1-AP protocol on the S1-MME interface are activated.

1.3.3. *RRC protocol*

The RRC protocol is the signaling exchanged between the mobile and the eNB entity over the LTE-Uu radio interface.

The RRC protocol performs the following functions:

– distribution of system information relating to the radio interface characteristics;

– control of the RRC connection: the procedure includes paging establishment, modification and release of the radio bearer assigned to Signaling Radio Bearer (SRB) and to the Data Radio Bearer (DRB). The procedure also includes security activation which consists of putting in place the mechanisms for encrypting traffic data and signaling and for signaling integrity control;

– control of handover: the procedure to perform the cell change between two entities eNB (intra-system handover) or between an eNB entity and a 2nd or 3rd generation mobile network entity (inter-system handover);

– measurement report: the eNB entity can start making measurements at the mobile periodically or upon request, to prepare the handover;

– transport of NAS messages between the mobile and the MME entity.

The mobile can be in two operational states, the idle state (RRC-IDLE) or the connected state (RRC-CONNECTED).

In the idle state, the mobile is unknown to the eNB entity and remains in this state until the completion of the procedure for establishing the RRC connection.

The transition to the connected state is initialized by the mobile when it needs to transmit traffic data or signaling.

In the connected state, the mobile is assigned a Cell Radio Network Temporary Identity (C-RNTI).

1.3.4. *Data link layer*

1.3.4.1. *PDCP protocol*

The PDCP protocol is used for RRC signaling messages, relating to dedicated control data, and for IP packet traffic data.

The PDCP protocol performs the following functions:

– compression of the traffic data headers using the Robust Header Compression (ROHC) mechanism;

– security of traffic data (confidentiality) and RRC signaling (integrity and confidentiality);

– delivery RRC messages and IP packets in sequences;

– recovery of PDCP frames lost during the handover.

Several PDCP instances can be activated simultaneously:

– two instances for SRB1 relating to RRC signaling data, and SRB2:

- the SRB1 bearer is used for the transmission of an RRC message that can carry a NAS message,

- the SRB2 bearer is used for the transmission of a NAS message only;

– one instance for each DRB radio bearer relating to traffic data.

1.3.4.2. *RLC protocol*

The RLC protocol provides control of the radio link between the mobile and the eNB entity.

The mobile can simultaneously activate multiple RLC instances, with each instance corresponding to a PDCP instance.

The RLC protocol operates in three modes:

– Acknowledged Mode (AM);

– Unacknowledged Mode (UM);

– Transparent Mode (TM) for which no header is added to data.

The RLC protocol performs the following operations:

– retransmission in the case of error via Automatic Repeat reQuest (ARQ) mechanism, for the acknowledged mode only;

– concatenation, segmentation and reassembly of PDCP frames both in the acknowledged and unacknowledged mode;

– possible resegmentation of PDCP frames, in the acknowledged mode, during retransmission of the RLC frame;

– resequencing of received data, both in the acknowledged and unacknowledged mode;

– detection of duplicate data both in the acknowledged and unacknowledged mode.

1.3.4.3. *MAC protocol*

The MAC protocol provides the following functions:

– multiplexing RLC frames from multiple instances in one or two transport blocks;

– resource allocation via a scheduling mechanism;

– management of retransmission in case of error via the Hybrid Automatic Repeat reQuest (HARQ) mechanism;

– management of the random access procedure.

1.3.5. *Logical channels*

The Broadcast Control Channel (BCCH) is a unidirectional common control channel, used only in the downlink direction for broadcasting of Master Information Block (MIB) and System Information Block (SIB) messages.

The Paging Control Channel (PCCH) is a unidirectional common control channel, used only in the downlink direction to transport RRC paging messages.

The Common Control Channel (CCCH) is a bidirectional common control channel, used to transmit the first RRC signaling messages when the mobile tries to connect to the eNB entity.

The Dedicated Control Channel (DCCH) is a bidirectional channel used to transmit RRC messages when the mobile is connected to the eNB entity.

The Dedicated Traffic Channel (DTCH) is a dedicated bidirectional channel, used to transmit unicast IP packets.

The Multicast Control Channel (MCCH) is a unidirectional channel used for transmitting RRC messages associated with IP packets transmitted in broadcast mode.

The Multicast Traffic Channel (MTCH) is a unidirectional channel used to transmit IP packets in broadcast mode to the mobile.

1.3.6. *Transport channels*

1.3.6.1. *Downlink direction*

The Broadcast Channel (BCH) supports the BCCH logical channel related to the MIB system information message.

The Paging Channel (PCH) supports the PCCH logical channel.

The Downlink Shared Channel (DL-SCH) multiplexes the CCCH, DCCH, DTCH logical channels and BCCH logical channel related to SIB system information messages.

The MCCH and MTCH are mapped to the DL-SCH transport channel if the number of mobiles concerned by IP packets transmitted in broadcast mode is low.

The Multicast Channel (MCH) multiplexes the MCCH and MTCH logical channels if the number of mobiles concerned by IP packets transmitted in broadcast mode is significant.

1.3.6.2. *Uplink direction*

The Random Access Channel (RACH) does not transport logical channels. It is used by the mobile for random access to the eNB entity. The RACH transport channel only carries a preamble to initialize the connection with the eNB entity.

The Uplink Shared Channel (UL-SCH) multiplexes the DCCH, CCCH and DTCH logical channels.

1.3.7. *Physical layer*

The transmission chain consists of two subsets:

– for each direction of transmission, the first subset comprises the error detection and error correction codes and the bit rate matching;

– for the downlink direction, the second subset includes the modulation, the mapping on the spatial layers, the precoding, the mapping on the resource elements and the Inverse Fast Fourier Transform (IFFT) to generate the Orthogonal Frequency Division Multiple Access (OFDMA) signal;

– for the uplink direction, the second subset includes the modulation, the mapping on the resource elements and the inverse fast Fourier transform. The generation of the Single Carrier Frequency Division Multiple Access (SC-FDMA) signal introduced a Fast Fourier Transform (FFT). The mapping on the spatial layers and the precoding are implemented only for Release 10.

The support for both transmission directions uses two bandwidths matched in the Frequency Division Duplex (FDD) mode or a single bandwidth in the Time Division Duplex (TDD) mode.

For the FDD mode, each transmission direction operates simultaneously in the assigned bandwidth.

For the TDD mode, both transmission directions operate alternately in the same bandwidth, meaning each direction is assigned a portion of time.

The bandwidth of the radio channel is flexible and can take several values: 1.4-MHz, 3-MHz, 5-MHz, 10-MHz, 15-MHz and 20-MHz.

Carrier Aggregation (CA) involves combining the use of several Component Carrier (CC) or radio channels to increase the cell bit rate. The aggregation can be performed on five radio channels, bringing the maximum bandwidth to 100-MHz.

The radio channel is formed in the frequency domain of an Orthogonal Frequency Division Multiplexing (OFDM) with subcarrier spacing of 15 kHz or 7.5 kHz.

The radio channel consists of 10-ms frames in the time domain, each frame comprising time slots, each time slot comprising OFDM symbols.

Four modes characterize the transmission system on the radio channel (Figure 1.12). It should be noted that the term *input* is applied to the input of the radio channel and the term *output* to the output of the same channel.

The Single Input Single Output (SISO) mode is the basic signal propagation mode in which a transmitting and a receiving antenna are used.

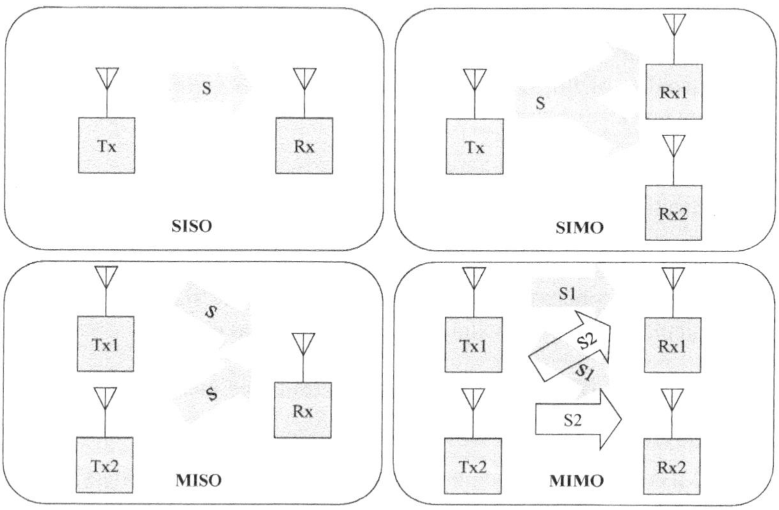

Figure 1.12. *Transmission modes*

The Single Input Multiple Output (SIMO) mode is characterized by the use of a single transmit antenna and multiple receiving antennas. The SIMO mode is often referred to as receive diversity. The transmitted bit rate is identical to the SISO mode. On the other side, the selection of the received signal allows the receiver to protect against the fading of the radio signal.

The Multiple Input Single Output (MISO) mode features multiple transmit antennas and a single receiving antenna. The same signal is transmitted on the transmit antennas. The MISO mode is often referred to as transmit diversity. As for the SIMO mode, the MISO mode allows the receiver to protect against the fading of the radio signal.

The MISO mode is also used to form beams (beamforming) directed towards the mobile, by controlling different phases of the transmitted signals.

The Multiple Input Multiple Output (MIMO) uses multiple antennas for transmission and reception. It improves bit rate by allowing the transmission of several different signals on the same frequency and at the same time.

1.3.8. *Physical signals*

1.3.8.1. *Downlink direction*

The Primary Synchronization Signal (PSS) ensures the frequency synchronization of the OFDMA signal and the time synchronization at the half-frame level.

The Secondary Synchronization Signal (SSS) provides the time synchronization at the frame level.

The Cell-Specific Reference Signal (RS) is a signal specific to the cell used to perform coherent demodulation of the received signal which is based on the calculation of the transfer function of the radio channel.

The Cell-Specific RS physical signal permits the measurement of RSRP and Reference Signal Received Quality (RSRQ) of the received signal.

The MBMS Single Frequency Network Reference Signal (MBSFN RS) is transmitted only in the Physical Multicast Channel (PMCH) to perform coherent demodulation of the received signal.

The UE-Specific RS physical signal is a specific signal to the mobile used to perform coherent demodulation of the received signal, to measure the power of the received signal and to form beams.

The Positioning Reference Signal (PRS) is used by the mobile to determine its location from the Observed Time Difference Of Arrival (OTDOA) mechanism.

The Channel State Information Reference Signal (CSI RS) improves the measurement of the received signal and the interference level from that supplied from the Cell-Specific RS physical signal.

The power of the CSI RS physical signal is either transmitted to determine the level of the received signal, or suppressed to measure the level of interference.

1.3.8.2. *Uplink direction*

The Demodulation Reference Signal (DM-RS) associated with the Physical Uplink Shared Channel (PUSCH) is used for estimation and synchronization of the PUSCH physical channel.

The DM-RS physical signal associated with the Physical Uplink Control Channel (PUCCH) is used for estimation and synchronization of the PUCCH physical channel.

The Sound Reference Signal (SRS) allows the eNB entity to measure the quality of the signal for the uplink direction, in a frequency band higher than that allocated to the mobile.

This measurement cannot be obtained by the DM-RS physical signal because the DM-RS is associated with the PUSCH or PUCCH physical channels.

The measurement performed by the eNB entity allows it to set the frequency location of the resources allocated to the mobile for the uplink direction, and the Modulation and Coding Scheme (MCS).

1.3.9. *Physical channels*

1.3.9.1. *Downlink direction*

The Physical Broadcast Channel (PBCH) transmits the BCH transport channel containing the system information corresponding to the MIB message.

The Physical Control Format Indicator Channel (PCFICH) transmits the Control Format Indicator (CFI) indicating the size of the Physical Downlink Control Channel (PDCCH).

The Physical HARQ Indicator Channel (PHICH) transmits the HARQ Indicator (HI) which indicates a positive ACK or negative NACK acknowledgment for the uplink data received in the Physical Uplink Shared Channel (PUSCH).

The PDCCH physical channel transmits Downlink Control Information (DCI):

– allocation of the resources and the modulation and coding scheme, for data contained in the PDSCH and PUSCH physical channels;

– transmission power of the PUCCH and PUSCH physical channels.

The PDSCH physical channel transmits the DL-SCH and PCH transport channels.

The Physical Multicast Channel (PMCH) transmits the MCH transport channel.

1.3.9.2. *Uplink direction*

The Physical Random Access Channel (PRACH) contains a preamble used by the mobile when it needs to perform a random access, which is the first stage of the connection of the mobile to the eNB entity.

The PUCCH physical channel uses three types of format to transport the Uplink Control Information (UCI):

– formats 1, 1a and 1b transport the UCI information relating to the scheduling request to obtain resource on the PUSCH physical channel and to the positive ACK or negative NACK acknowledgment, corresponding to the HARQ mechanism, for the data received on the PDSCH physical channel;

– formats 2, 2a and 2b transport UCI information relating to the signal status reports for the signal received on the PDSCH physical channel and to the positive ACK or negative NACK acknowledgments;

– format 3 transports the same information as format 1 by adapting it to the aggregation of radio channels introduced in Release 10.

The PUSCH physical channel transmits the UL-SCH transport channel and the UCI control information.

For Release 8 and 9, the simultaneous transmission of PUSCH and PUCCH physical channels is not supported.

Transmitting UCI information in the PUSCH physical channel is carried out, on the one hand, together with the transfer of traffic data or RRC control, and on the other hand, for the transfer of a periodic UCI information reports.

For Release 10, the simultaneous transmission of the PUSCH and PUCCH physical channels is supported.

The transmission of UCI information in the PUCCH physical channel is maintained when the traffic data or RRC control need to be transferred to the PUSCH physical channel.

1.3.10. *Mobile categories*

The mobile categories determine the maximum bit rate on the LTE-Uu radio interface for the downlink and the uplink direction (Table 1.4).

The maximum bit rate depends on the optimum characteristics of the radio interface (modulation, radio channel bandwidth, MIMO mechanism) and the ability of the mobile to treat the bit rate allowed by the radio conditions.

Mobiles of categories 1 to 5 are LTE mobiles defined in Release 8.

Mobiles of categories 6 to 8 are LTE Advanced mobiles defined in Release 10.

Given the 4×4 MIMO processing difficulty for mobiles of category 5, mobiles of categories 6 and 7 were introduced to provide the 300 Mbps bit rate for the downlink direction, obtained by keeping the 2×2 MIMO and doubling the bandwidth of the radio channel.

Category 7 mobiles allow the 75 Mbps bit rate for mobiles of category 5 to be exceeded for the uplink direction, by doubling the bandwidth of the radio channel and avoiding the use of 64-Quadrature Amplitude Modulation (64-QAM).

Radio interface	LTE					LTE Advanced		
Category	1	2	3	4	5	6	7	8
DL bit rate Mbps	10	50	100	150	300	300	300	3000
UL bit rate Mbps	5	25	50	50	75	50	100	1500
Bandwidth MHz	20	20	20	20	20	2X20 DL	2X20 DL UL	5X20 DL UL
Modulation DL	64 QAM	64 QAM	64 QAM	64 QAM	64 QAM	64 QAM	64 QAM	64 QAM
Modulation UL	16 QAM	16 QAM	16 QAM	16 QAM	64 QAM	16 QAM	16 QAM	64 QAM
MIMO DL	N.A.	2X2	2X2	2X2	4X4	2X2	2X2	8X8
MIMO UL	N.A.	N.A.	N.A.	N.A.	N.A.	N.A.	N.A.	4X4

DL: (*Downlink*)
UL: (*Uplink*)

Table 1.4. *Mobile categories*

1.4. Network procedures

1.4.1. *Connection procedure*

The connection procedure is initiated by the mobile in various circumstances:

– the user turns on his mobile and attaches to a network;

– the mobile is idle and it has to update its location;

– the mobile is idle and it wants to establish a session;

– the mobile is idle and incoming data is intended for it;

– the mobile makes a handover during a session.

The connection procedure is preceded by the random access procedure described in Chapter 4.

When establishing the connection, the RRC *ConnectionRequest* message, which uses the Signaling Radio Bearer 0 (SRB0) of the Common Control Channel (CCCH), is transmitted to the evolved Node B (eNB) entity.

The eNB entity responds to the mobile by the RRC *ConnectionSetup* message that provides the Logical Channel Identifier (LCID) of the Dedicated Control Channel (DCCH) and the SRB1 bearer parameters.

The mobile confirms the establishment of the SRB1 bearer of the DCCH logical channel by sending the RRC *ConnectionSetupComplete* message.

1.4.2. *Attachment procedure*

At the end of the connection procedure, the mobile starts the attachment procedure which comprises the following steps:

– the mutual authentication between the User Equipment (UE) and the Mobility Management Entity (MME) corresponding to the Authentication and Key Agreement (AKA) mechanism;

– the security of Non-Access Stratum (NAS) messages;

– the registration of the Tracking Area Identity (TAI) to the MME entity;

– the establishment of a default bearer;

– the granting of a Globally Unique Temporary Identity (GUTI).

The mobile attachment procedure is described in Figures 1.13 and 1.14.

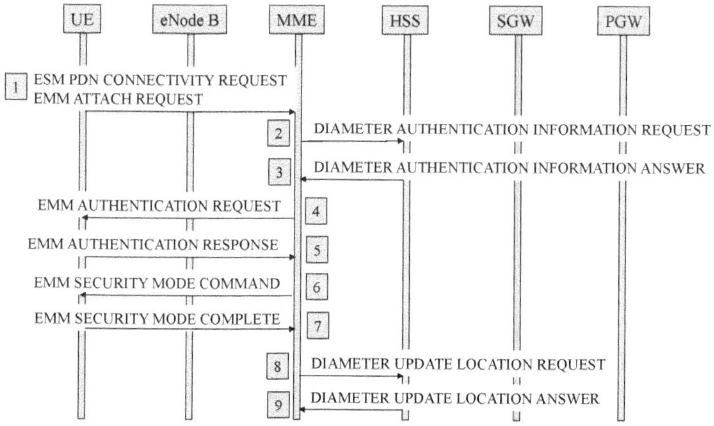

Figure 1.13. *Mobile attachment procedure: NAS message authentication and security*

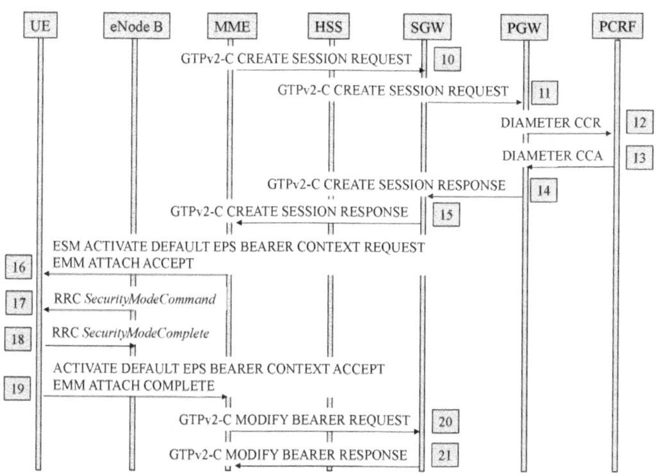

Figure 1.14. *Mobile attachment procedure: establishment of the default bearer and of the security parameter of the radio interface*

1) The attachment procedure is triggered by the mobile when it sends the MME entity the EMM ATTACH REQUEST message containing the International Mobile Subscriber Identity (IMSI).

The EMM ATTACH message carries the message ESM PDN CONNECTIVITY REQUEST.

The EMM ATTACH REQUEST message is carried by the RRC *ConnectionSetupComplete* message on the LTE-Uu radio interface and by the S1-AP INITIAL UE MESSAGE on the S1-MME interface.

The S1-AP UE INITIAL MESSAGE contains the TAI location area of the mobile and the E-UTRAN Cell Global Identifier (ECGI).

2) Upon receipt of the EMM ATTACH REQUEST message, the MME entity requests the Home Subscriber Server (HSS) entity for the cryptographic data of the mobile in the DIAMETER AUTHENTICATION INFORMATION REQUEST message.

3) The HSS entity generates cryptographic data using a random number (RAND) and the Ki key of the mobile and sends them to the MME entity in the DIAMETER AUTHENTICATION INFORMATION ANSWER message.

The cryptographic data contain the random number, the mobile RES authentication code, the AUTN network authentication code and the K_{ASME} key.

The MME entity derives the K_{ASME} key to generate the CK_{NAS}, IK_{NAS} and K_{eNB} keys:

- the CK_{NAS} key is used to encrypt the NAS message;
- the IK_{NAS} key is used to control the integrity of the NAS message;
- the K_{eNB} key is transferred to the eNB entity.

4) The MME entity transmits the EMM AUTHENTICATION REQUEST message containing the random number and the AUTN network authentication code to the mobile.

The mobile calculates locally, from its Ki key stored in the Universal Services Identity Module (USIM) of its Universal Integrated Circuit Card (UICC) and from the random number received, the K_{ASME} key, its RES authentication code and that of the AUTN network authentication code which it compares to the value received. If the two values are identical, the network is authenticated.

5) The mobile responds to the MME entity with the EMM AUTHENTICATION RESPONSE message containing the RES authentication code.

The MME entity compares the RES values received from the mobile and from the HSS entity. If the two values are the same, the mobile is authenticated.

6) The security parameters for the NAS signaling are enabled by the MME entity sending the EMM SECURITY MODE COMMAND message controlled with the IK_{NAS} key. This message contains algorithms to derive the K_{ASME} key.

The mobile derives the K_{ASME} key to generate the CK_{NAS}, IK_{NAS}, and K_{eNB} keys and checks the integrity of the message EMM SECURITY MODE COMMAND.

7) The mobile responds with the message EMM SECURITY MODE COMPLETE encrypted with the CK_{NAS} key and controlled with IK_{NAS} key.

After the mutual authentication phase and the securing of NAS messages, the MME entity registers the mobile with the HSS entity.

8) The MME sends the DIAMETER LOCATION UPDATE REQUEST message to the HSS entity to register the mobile and obtain its profile.

The HSS entity registers the identity of the MME entity which attached the mobile and the TAI location area.

9) The HSS entity responds to the MME entity with the DIAMETER ANSWER LOCATION UPDATE message that contains the profile of the mobile:

– the Access Point Names (APN);

– the Quality of Service (QoS) characteristics for each default bearer that must be established.

The MME entity selects the Serving Gateway (SGW) entity in its group and the PDN Gateway (PGW) entity from a Domain Name Server (DNS) resolution of the APN.

10) The MME entity sends the GTPv2-C CREATE SESSION REQUEST message to create a context at the SGW entity.

The GTPv2-C CREATE SESSION REQUEST message contains the Internet Protocol (IP) address of the PGW entity, the APN and the default bearer profile.

11) The SGW entity sends the GTPv2-C CREATE SESSION REQUEST message to create a context at the PGW entity.

The GTPv2-C CREATE SESSION REQUEST message contains the Tunnel Endpoint Identifier (TEID) that the PGW entity will use at the GPRS Tunneling Protocol User (GTP-U) protocol level for the S5 bearer.

12) The PGW entity sends the Policy and Charging Rules Function (PCRF) the DIAMETER CCR (CREDIT-CONTROL REQUEST) message for authorization to open the default bearer.

The PCRF entity compares the profile of the mobile with the rules defined for the network and stored in the Subscription Profile Repository (SPR) entity.

13) The PCRF entity responds to the PGW entity by a DIAMETER CCA (CREDIT CONTROL-ANSWER) message containing the rules to be applied to the default bearer (filter parameters, charging mode).

14) The PGW entity responds to the SGW entity by a GTPv2-C CREATE SESSION RESPONSE message which contains the following information:

– the TEID identifier that the SGW entity will use at the GTP-U protocol level for the S5 bearer;

– the mobile configuration (IP address of the mobile, IP address of its DNS server).

15) The SGW entity responds to the MME entity with a GTPv2-C CREATE SESSION RESPONSE which contains the following information:

– the TEID identifier that the eNB entity will use at the GTP-U protocol level for the S1 bearer;

– the mobile configuration.

16) The MME entity responds to the mobile with the EMM ATTACH ACCEPT message containing the following information:

– the mobile configuration;

– the GUTI temporary identity.

The EMM ATTACH ACCEPT message carries the ESM ACTIVATE DEFAULT EPS BEARER CONTEXT REQUEST message.

The EMM ATTACH ACCEPT message is carried by the S1-AP INITIAL CONTEXT SETUP REQUEST message on the S1-MME interface and by the RRC *ConnectionReconfiguration* message on the LTE-Uu radio interface.

The S1-AP INITIAL CONTEXT SETUP REQUEST message allows the creation of the mobile context at the eNB entity level and contains the following information:

– the TEID identifier assigned by the SGW entity;

– the QoS parameters;

– the K_{eNB} key derived from the K_{ASME} key.

The eNB entity derives the K_{eNB} key for creating the following keys:

– the K_{RRCenc} key for RRC message encryption;

– the K_{RRCint} key for RRC message integrity control;

– the K_{UPenc} traffic data (IP packets) encryption key.

The RRC *ConnectionReconfiguration* message initializes the mounting of the DRB radio bearer.

17) The eNB entity requests the mobile to secure the radio interface with the RRC *SecurityModeCommand* message which is controlled with the K_{RRCint} integrity key and contains algorithms that allow the mobile to derive the K_{eNB} key.

The mobile derives the K_{eNB} key to generate the K_{RRCenc}, K_{RRCint}, and K_{UPenc} keys and checks the integrity of the RRC *SecurityModeCommand* message.

18) The mobile confirms the establishment of the keys to the eNB entity with the RRC *SecurityModeComplete* message which is controlled with the integrity key K_{RRCint}.

The messages of steps 17 and 18 are inserted between the reception of the S1-AP INITIAL CONTEXT SETUP REQUEST message and the transmission of the RRC *ConnectionReconfiguration* message by the eNB entity.

19) The mobile confirms receipt of the EMM ATTACH REQUEST message by sending the EMM ATTACH COMPLETE message.

The EMM ATTACH COMPLETE message carries the ESM ACTIVATE DEFAULT EPS BEARER CONTEXT ACCEPT message.

The EMM ATTACH COMPLETE message is carried by the RRC *ConnectionReconfigurationComplete* message on the LTE-Uu radio interface and by the S1-AP INITIAL CONTEXT SETUP RESPONSE message on the S1-MME interface.

The S1-AP INITIAL CONTEXT SETUP RESPONSE message contains the TEID identifier which the SGW entity will use at the GTP-U protocol level for the S1 bearer.

20) The MME entity transfers the TEID identifier received from the eNB entity to the SGW entity in the GTPv2-C MODIFY BEARER REQUEST message.

21) The SGW entity acknowledges the message received by the GTPv2-C MODIFY BEARER RESPONSE message.

1.4.3. *Restoration procedure of the default bearer*

When the mobile is idle, the restoration of the default bearer can be activated by the mobile in the case of outgoing data or by the SGW entity in the case of incoming data.

1.4.3.1. *Default bearer restoration initiated by the mobile*

The restoration procedure of the default bearer, initiated by the mobile, is described in Figure 1.15.

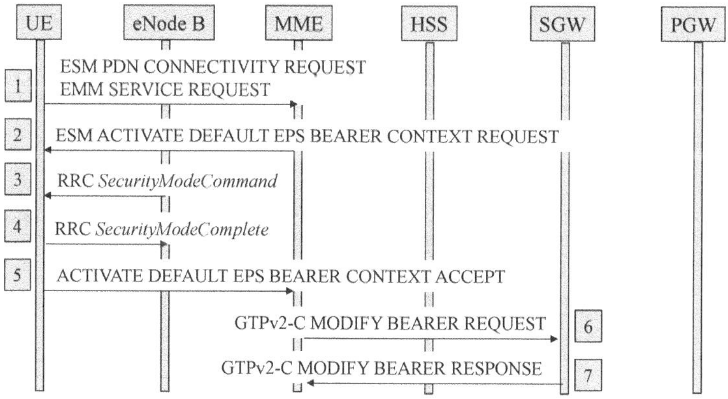

Figure 1.15. *Restoration procedure of the default bearer initiated by the mobile*

1) In the case of outgoing data, the mobile transmits the EMM SERVICE REQUEST message to the MME entity.

The EMM SERVICE REQUEST message is carried by the RRC *ConnectionSetupComplete* message on the LTE-Uu radio interface and by the S1-AP INITIAL MESSAGE UE message on the S1-MME interface.

The EMM SERVICE REQUEST message carries the ESM PDN CONNECTIVITY REQUEST message.

The EMM SERVICE REQUEST message integrity is controlled with the IK_{NAS} key.

The MME entity retains the security parameters stored in the context if the integrity checking is positive.

Otherwise, the MME entity starts a new AKA procedure.

2) The MME entity transmits the ESM ACTIVATE DEFAULT EPS BEARER CONTEXT REQUEST message to the mobile.

The message ESM ACTIVATE DEFAULT EPS BEARER CONTEXT REQUEST is carried by:

– the S1-AP INITIAL CONTEXT SETUP REQUEST message to create the context of the mobile at the eNB entity level;

– the RRC *ConnectionReconfiguration* message for the mounting the DRB radio bearer.

3) The eNB entity requests the mobile to secure the radio interface with the RRC *SecurityModeCommand* message.

4) The mobile confirms the establishment of the keys to the eNB entity with the RRC *SecurityModeComplete* message.

5) The mobile responds to the MME by the message ACTIVATE DEFAULT EPS BEARER ESM CONTEXT ACCEPT carried by:

– the RRC *ConnectionReconfigurationComplete* message;

– the S1-AP INITIAL CONTEXT SETUP RESPONSE message containing the TEID identifier that SGW entity will use the GTP-U protocol level for the S1 bearer.

6) The MME entity transfers the TEID identifier received from the eNB entity to the SGW entity in the GTPv2-C MODIFY BEARER REQUEST message.

7) The SGW entity acknowledges the message received by the GTPv2-C MODIFY BEARER RESPONSE message.

1.4.3.2. Default bearer restoration initiated by the SGW entity

The default bearer restoration procedure initiated by the SGW entity is described in Figure 1.16.

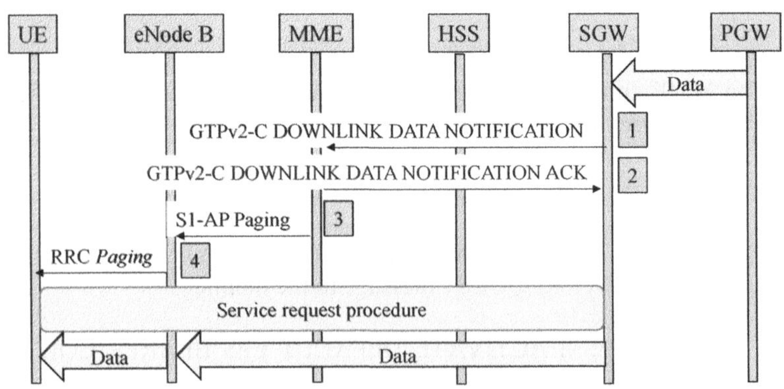

Figure 1.16. *Restoration procedure of the default bearer initiated by the SGW entity*

1) The SGW entity notifies the MME entity of incoming data reception by the GTPv2-C DATA DOWNLINK NOTIFICATION message.

2) The MME entity transmits the GTPv2-C NOTIFICATION DATA DOWNLINK ACK message to the SGW entity to acknowledge the received message.

3) The MME transmits thc S1-AP PAGING message, containing the S-TMSI identity of the mobile, to all eNB entities of the TAI location area. The S-TMSI is a part of the GUTI temporary identity.

4) The RRC *Paging* message containing the S-TMSI identity of the mobile is broadcast in the cell by each eNB entity of the TAI location area.

On receipt of the RRC paging message, the mobile starts the service request procedure described in the previous section.

1.4.4. Establishment procedure of a dedicated bearer

The establishment of a dedicated bearer, assigned for example to the voice, is coupled with the establishment of a default bearer, assigned for example to the telephone signaling.

The establishment of the dedicated bearer is triggered by the Application Function (AF) entity, for example by the IP Multimedia Subsystem (IMS) network, from the analysis of telephone signaling exchanged between the terminals that want to establish a telephone call.

The link between the IMS network and the EPS network is provided by the PCRF entity.

The PCRF entity transfers the characteristics of the dedicated bearer assigned to the voice that needs to be established to the PGW entity.

This dedicated bearer is coupled with the default bearer assigned to telephone signaling, in that the bearer terminations are the same for both types of bearer.

The procedure for establishing dedicated bearers is described in Figure 1.17.

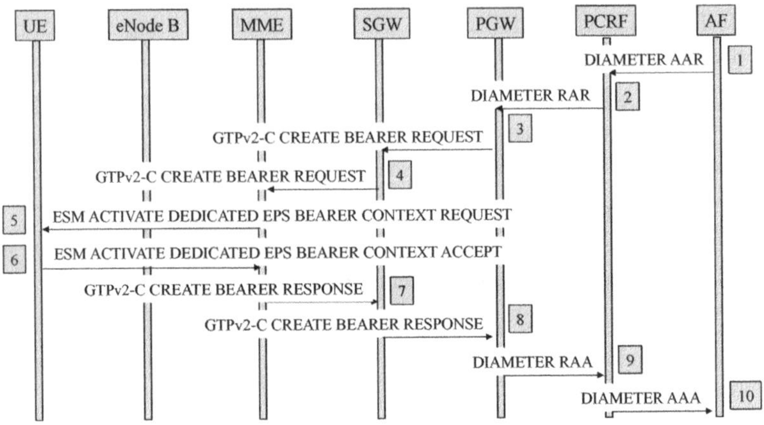

Figure 1.17. *Procedure for establishing a dedicated bearer*

1) The AF entity sends the PCRF entity the dedicated bearer characteristics in the DIAMETER AAR (Authenticate Authorize Request) message.

2) The PCRF entity sends the PGW entity the dedicated bearer characteristics in the DIAMETER RAR (Re-Auth Request) message.

3) The PGW entity sends the SGW entity the GTPv2-C CREATE BEARER REQUEST message containing the TEID identifier of the S5 bearer that the SGW entity will need to use in the GTP-U header when sending traffic data to the PGW entity.

4) The SGW entity sends the MME entity the GTPv2-C CREATE BEARER REQUEST message containing the TEID identifier of the S1 bearer that the eNB entity will need to use in the GTP-U header when sending traffic data to the SGW entity;

5) The MME entity sends the mobile the ESM ACTIVATE DEDICATED EPS BEARER CONTEXT REQUEST message carried by:

– the S1-AP E-RAB SETUP REQUEST message on the S1-MME interface;

– the RRC *ConnectionReconfiguration* message on the LTE-Uu interface.

The S1-AP E-RAB SETUP REQUEST message contains the TEID identifier that the MME entity has received from the SGW entity.

The RRC *ConnectionReconfiguration* message contains the LCID identifier of the dedicated bearer.

6) The mobile responds to the MME entity by the ESM ACTIVATE DEDICATED EPS BEARER CONTEXT ACCEPT message carried by:

– the RRC *ConnectionReconfigurationComplete* message on the LTE-Uu interface;

– the S1-AP E-RAB SETUP RESPONSE message on the S1-MME interface.

The S1-AP E-RAB SETUP RESPONSE message contains the TEID identifier of the S1 bearer that the SGW entity will need to use in the GTP-U header when it sends traffic data to the eNB entity.

7) The MME entity responds to the SGW entity by the GTPv2-C CREATE BEARER RESPONSE message containing the TEID identifier that the MME entity has received from the eNB entity.

8) The SGW entity reponds to the PGW entity with the C-GTPv2 CREATE BEARER RESPONSE message containing the TEID identifier of the S5 bearer that PGW entity will need to use in the GTP-U header when it sends traffic data to the SGW entity.

9) The PGW entity responds to the PCRF entity with the DIAMETER RAA (Re-Auth Answer) message.

10) The PCRF entity responds to the AF entity with the DIAMETER AAA message (Authenticate Authorize Answer).

1.4.5. *Location update procedure*

The location update procedure is initiated by the mobile in idle mode when entering a new TAI location area or when its maintenance timer expires.

The location update procedure is described in Figure 1.18.

The mobile runs the connection process before starting the location update.

1) The mobile sends the EMM TRACKING AREA UPDATE REQUEST message indicating the cause of the location update.

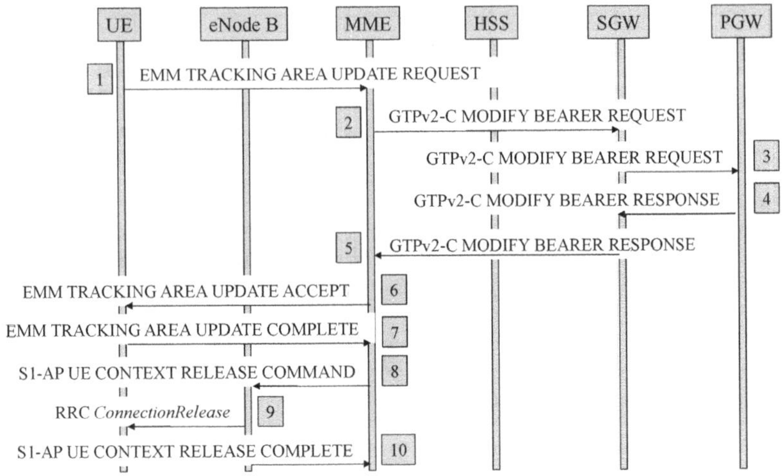

Figure 1.18. *Location update procedure*

The EMM TRACKING AREA UPDATE REQUEST message is carried by the RRC *ConnectionSetupComplete* message on the radio LTE-Uu interface and by the S1-AP INITIAL MESSAGE UE message on the S1-MME interface.

The RRC *ConnectionSetupComplete* message contains the Globally Unique MME Identity (GUMMEI) that the eNB entity uses to transfer the EMM TRACKING AREA UPDATE REQUEST message to the MME entity.

The integrity of the EMM TRACKING AREA UPDATE REQUEST message is controlled with the IK_{NAS} key.

The MME entity retains the security parameters stored in the context if the integrity checking is positive.

Otherwise, the MME entity starts a new AKA procedure.

2) The MME entity sends the SGW entity the GTPv2-C MODIFY BEARER REQUEST message containing the identities ECGI of the cell and TAI of the location area.

3) The SGW entity checks if the identities have been changed, and if so, sends the GTPv2-C MODIFY BEARER REQUEST message to the PGW entity to inform it of this change.

4) The PGW entity responds to the SGW entity by the GTPv2-C MODIFY BEARER RESPONSE message.

5) The SGW entity responds to the MME entity by the GTPv2-C MODIFY BEARER RESPONSE message.

6) The MME can configure a new TAI location area list and assign a new GUTI identity to the mobile in the EMM TRACKING AREA UPDATE ACCEPT message.

The EMM TRACKING AREA UPDATE ACCEPT message is carried by the S1-AP DOWNLINK NAS TRANSPORT message on the S1-MME interface and by the RRC *ConnectionSetupComplete* message on the LTE-Uu radio interface:

7) If the GUTI identity was changed, the mobile sends the MME entity the EMM TRACKING AREA UPDATE COMPLETE message to acknowledge the received message.

8) The MME entity requests the eNB entity to disconnect the mobile in an S1-AP UE CONTEXT RELEASE COMMAND message.

9) The eNB entity disconnects the mobile by sending it the RRC *ConnectionRelease* message.

10) The eNB entity informs the MME entity of the disconnection of the mobile in the S1-AP UE CONTEXT RELEASE COMPLETE message.

1.4.6. *Handover procedure*

1.4.6.1. *Handover based on X2*

The handover based on X2 allows the exchange of commands directly between the eNB entities belonging to the same group.

The handover based on X2 is performed on the condition that a new MME entity is not selected.

The handover procedure based on X2 is described in Figure 1.19.

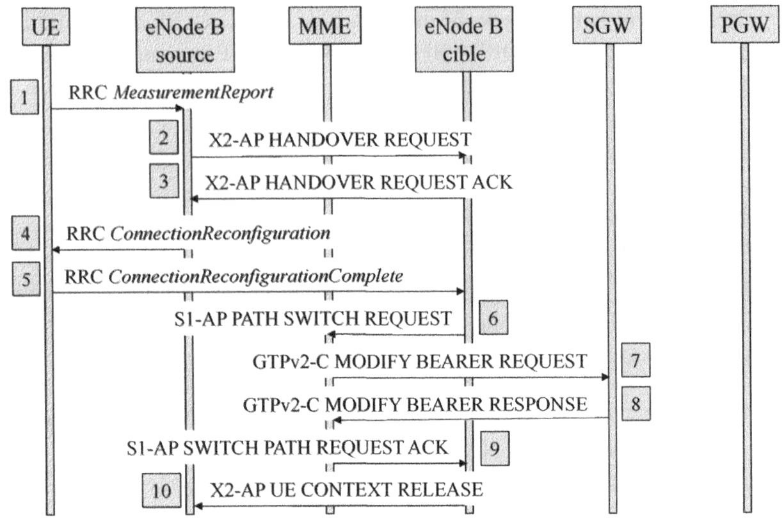

Figure 1.19. *Handover procedure based on X2*

1) The mobile performs measurements on its cell and on the neighboring cells, and transmits the results to the eNB entity in the RRC *MeasurementReport* message.

2) The source eNB entity selects the target eNB entity and transmits the X2-AP HANDOVER REQUEST message that contains the context of the mobile.

3) The target eNB entity reserves resources and responds to the source eNB entity by the X2-AP HANDOVER REQUEST ACK message containing the *Handover Command* information element and the TEID identifier of the X2-U bearer.

The *Handover Command* information element contains the characteristics of the radio interface that will enable the mobile to perform the handover in a few tens of milliseconds.

The X2-U bearer is an unidirectional bearer, from the source eNB entity to the target eNB entity.

The X2-U bearer is a temporary bearer which allows the source eNB entity to transfer the incoming data to the target eNB entity, when the mobile will be disconnected during the handover phase.

4) The source eNB entity instructs the mobile to change cells by an RRC *ConnectionReconfiguration* message containing the *HandoverCommand* information element.

5) The connection to the target eNB entity is finalized when the mobile sends the target eNB entity the RRC *ConnectionReconfigurationComplete* message.

6) The target eNB entity notifies the MME entity of the cell change by the S1-AP PATH SWITCH REQUEST message that contains the TEID identifier for the S1 bearer that needs to be changed.

7) The MME entity transfers the TEID identifier received from the eNB entity to the SGW entity in the GTPv2-C MODIFY BEARER REQUEST message.

8) The SGW entity acknowledges the message received by the GTPv2-C MODIFY BEARER RESPONSE message.

9) The MME acknowledges the message received from the target eNB entity by the S1-AP SWITCH PATH REQUEST ACK message.

10) The target eNB entity informs the source eNB entity that the context of the mobile can be removed by the X2-AP UE CONTEXT RELEASE message.

1.4.6.2. *Handover based on S1*

The handover based on S1 allows exchange of commands via the MME entity for the following cases:

– the X2 interface is not enabled between the eNB entities belonging to the same group;

– the two eNB entities belong to two different groups and new MME and SGW entities must be selected.

The RRC messages exchanged over the radio interface, however, remain identical to those of the handover procedure based on X2 interface.

1.4.7. *Multicast bearer establishment procedure*

Multicast bearer establishment initializes the transmission of a new Multimedia Broadcast Multicast Service (MBMS) session.

The procedure for establishing the multicast bearer is described in Figure 1.20.

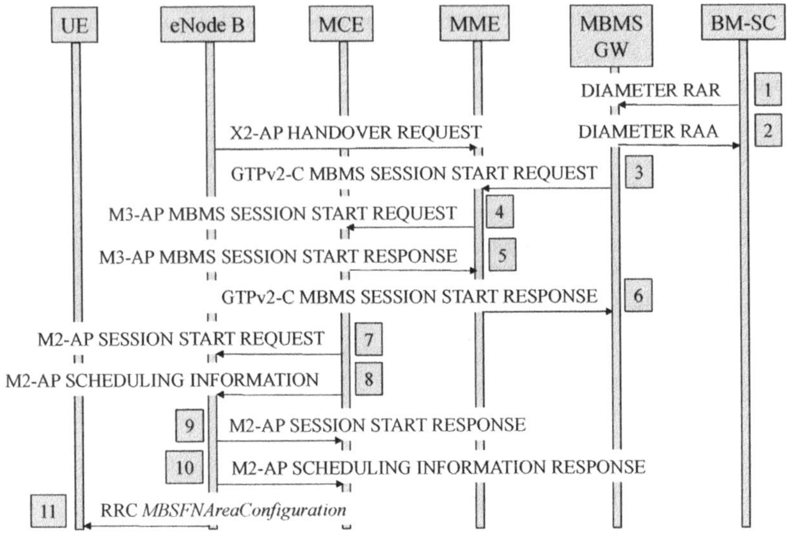

Figure 1.20. *Multicast bearer establishment procedure*

1) The procedure is initiated by the Broadcast Multicast Service Centre (BM-SC) entity by sending, to the MBMS Gateway (GW) entity, the DIAMETER RAR (Re-Auth Request) message containing the characteristics of the bearer that must be created.

2) The MBMS GW entity responds to the BM-SC entity by the DIAMETER RAA (Re-Auth Answer) message.

3) The MBMS GW entity initiates the construction of the multicast bearer by sending, to the MME entity, the GTPv2-C MBMS SESSION START REQUEST message, containing the multicast IP address of the bearer to create.

4) The MME entity distributes the M3-AP MBMS SESSION START REQUEST message to the Multi-cell/Multicast Coordination Entity (MCE) entities, containing the multicast IP address of the bearer to create.

5) Each MCE entity responds to the MME entity by the M3-AP MBMS SESSION START RESPONSE message.

6) The MME entity responds to the MBMS GW entity by the GTPv2-C MBMS SESSION START RESPONSE message.

7) The MCE entity sends the relevant eNB entities the M2-AP START SESSION REQUEST message containing the multicast IP address of the bearer to create.

8) The MCE entity simultaneously sends the relevant eNB entities the M2-AP SCHEDULING INFORMATION message that defines the characteristics of the Multicast Traffic Channel (MTCH).

9) and 10) The eNB entity responds to both messages received by the M2-AP SESSION START RESPONSE and M2-AP SCHEDULING INFORMATION RESPONSE messages.

The eNB entities send the IP transport network the IGMP JOIN message to receive the multicast bearer.

11) The eNB entity transmits the MTCH logical channel characteristics to the mobiles in an RRC *MBSFNAreaConfiguration* message.

2

NAS Protocol

2.1. Attachment

2.1.1. *Procedure*

The attachment procedure is described in Figure 2.1.

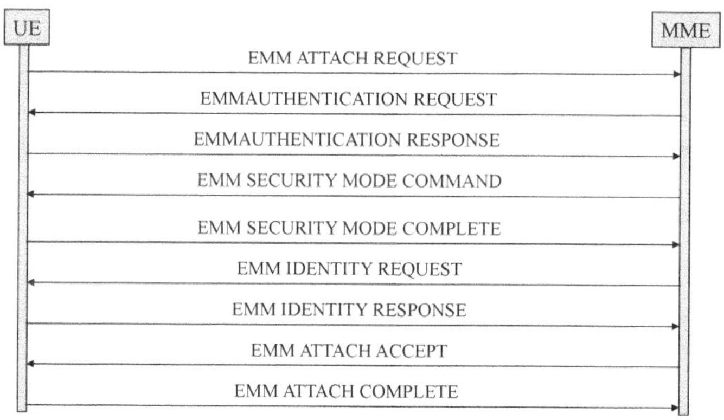

Figure 2.1. *Attachment procedure*

The attachment procedure is initiated by the mobile in the EMM-DEREGISTRERED state, by sending the EMM ATTACH REQUEST message to the Mobility Management Entity (MME).

This message contains the Globally Unique Temporary Identity (GUTI) or the International Mobile Subscriber Identity (IMSI).

On receipt of this message, the MME entity starts the authentication procedures and the security of Non-Access Stratum (NAS) messages.

If successful, the MME entity responds with the EMM ATTACH ACCEPT message containing a new GUTI identity.

The mobile switches to the EMM-REGISTERED state and responds with the EMM ATTACH COMPLETE message to acknowledge the previous message.

If the procedure fails, the MME entity responds with the EMM ATTACH REJECT message, causing the detachment of the mobile.

Detachment can be triggered by the mobile or the MME entity by sending the EMM DETACH REQUEST message.

The EMM DETACH ACCEPT response concludes the detachment process. The response is not transmitted by the MME entity when the detachment initialized by the mobile indicates a switch-off.

The mutual authentication procedure is initiated by the MME entity by sending the EMM AUTHENTICATION REQUEST message containing a random number (RAND) and the Authentication Network (AUTN) code.

The mobile locally calculates its RES authentication code and that of the network from the RAND received which it compares to the AUTN authentication code received from the MME entity.

If the MME entity is authenticated, the mobile responds with the EMM AUTHENTICATION RESPONSE message containing the mobile RES authentication code.

Otherwise, it indicates that network authentication has failed in the EMM AUTHENTICATION FAILURE message.

The MME entity compares the RES authentication code received by the mobile with the RES authentication code communicated by the Home Subscriber Server (HSS) entity. If the two authentication codes are identical, the mobile is authenticated and the MME entity triggers NAS signaling security. Otherwise, it responds with the EMM AUTHENTICATION REJECT message.

When mutual authentication is successful, the MME entity causes security implementation of the NAS signaling by sending the EMM SECURITY MODE COMMAND message. This message is integrity protected.

If the integrity control of the EMM SECURITY MODE COMMAND message is positive, the mobile responds with the EMM COMPLETE SECURITY MODE message. All subsequent NAS messages are encrypted and integrity controlled.

If the integrity control of the EMM SECURITY MODE COMMAND message is negative, the mobile responds with the EMM SECURITY MODE REJECT message.

When the NAS messages are secured, the MME entity requests the mobile for the International Mobile Equipment Identity (IMEI) of its device in an EMM IDENTITY REQUEST message.

The mobile provides the requested identity in the EMM IDENTITY RESPONSE message.

2.1.2. *Message structure*

2.1.2.1. *Attachment*

Table 2.1 presents the main information elements of the EMM ATTACH REQUEST message.

Information elements	Description
EPS attach type	EPS, EPS/IMSI, EPS emergency
NAS key set identifier	Security context identifier
EPS mobile identity	GUTI, IMSI
EU network capability	Security algorithms, SRVCC, CSFB, LPP, LCS
ESM message container	ESM PDN CONNECTIVITY REQUEST message

Table 2.1. *Information elements of the EMM ATTACH REQUEST message*

Table 2.2 presents the main information elements of the EMM ATTACH ACCEPT message.

Information elements	Description
EPS attach result	EPS, EPS / IMSI
T3412 value	Location update timer
TAI list	List of location areas
ESM message container	ESM ACTIVATE DEFAULT EPS BEARER CONTEXT REQUEST message
GUTI	Temporary mobile identity

Table 2.2. *Information elements of the EMM ATTACH ACCEPT message*

Table 2.3 presents the main information elements of the EMM ATTACH COMPLETE message.

Information elements	Description
ESM message container	ESM ACTIVATE DEFAULT EPS BEARER CONTEXT ACCEPT message

Table 2.3. *Information elements of the EMM ATTACH COMPLETE message*

Table 2.4 presents the main information elements of the EMM ATTACH REJECT message.

Information elements	Description
EMM cause	Attachment reject cause
T3402 value	Timer when resetting an attachment

Table 2.4. *Information elements of the EMM ATTACH REJECT message*

Table 2.5 presents the main information elements of the EMM DETACH REQUEST message.

Information elements	Description
Detach type	EPS, EPS/IMSI, IMSI
NAS key set identifier	Security context identifier
EPS mobile identity	GUTI, IMSI

Table 2.5. *Information elements of the EMM DETACH REQUEST message*

The EMM DETACH ACCEPT message contains no information element.

2.1.2.2. *Authentication*

Table 2.6 presents the main information elements of the EMM AUTHENTICATION REQUEST message.

Information elements	Description
NAS key set identifier	Security context identifier
Authentication parameter RAND	Random value
Authentication parameter AUTN	Network authentication code

Table 2.6. *Information elements of the EMM AUTHENTICATION REQUEST message*

Table 2.7 presents the main information elements of the EMM AUTHENTICATION RESPONSE message.

Information elements	Description
Authentication response parameter	RES authentication code

Table 2.7. *Information elements of the EMM AUTHENTICATION RESPONSE message*

Table 2.8 presents the main information elements of the EMM AUTHENTICATION FAILURE message.

Information elements	Description
EMM cause	Authentication reject cause

Table 2.8. *Information elements of the EMM AUTHENTICATION FAILURE message*

The EMM AUTHENTICATION REJECT message contains no information elements.

2.1.2.3. *Security mode control*

Table 2.9 presents the main information elements of the EMM SECURITY MODE COMMAND message.

Information elements	Description
Selected NAS security algorithms	integrity control and encryption algorithms for NAS message
NAS key set identifier	Security context identifier
Replayed UE security capabilities	Non-alteration control of the mobile security capabilities

Table 2.9. *Information elements of the EMM SECURITY MODE COMMAND message*

The EMM SECURITY MODE COMPLETE message contains no information elements

Table 2.10 presents the main information elements of the EMM SECURITY MODE REJECT message.

Information elements	Description
EMM cause	Security mode reject cause

Table 2.10. *Information elements of the EMM SECURITY MODE REJECT message*

2.1.2.4. *Identification*

Table 2.11 presents the main information elements of the EMM IDENTITY REQUEST message.

Information elements	Description
Identity Type	Request related to the IMEI identifier
Spare half octet	Filling

Table 2.11. *Information elements of the EMM IDENTITY REQUEST message*

Table 2.12 presents the main information elements of the EMM IDENTITY RESPONSE message.

Information elements	Description
Mobile identity	Value of the IMEI identifier

Table 2.12. *Information elements of the EMM IDENTITY RESPONSE message*

2.2. Session establishment

2.2.1. *Procedure*

The establishment of the default bearer occurs when attaching the mobile which simultaneously transmits the EMM ATTACH REQUEST and ESM PDN CONNECTIVITY REQUEST messages.

When the attachment of the mobile is accepted by the MME entity, the mobile simultaneously receives EMM ATTACH ACCEPT and ESM ACTIVATE DEFAULT EPS BEARER CONTEXT REQUEST messages.

The mobile acknowledges the two messages received by responding simultaneously EMM ATTACH COMPLETE and ESM ACTIVATE DEFAULT EPS BEARER CONTEXT ACCEPT.

When the mobile is in the RRC-IDLE state, default bearer restoration can be triggered by either the mobile, for outgoing data, or by the network, for incoming data.

When the mobile triggers default bearer restoration, it initializes the service request procedure by transmitting the EMM SERVICE REQUEST message to the MME entity.

The rest of the procedure repeats the messages exchanged previously:

– ACTIVATE DEFAULT EPS BEARER CONTEXT REQUEST message, from the MME entity to the mobile;

– ESM ACTIVATE DEFAULT EPS BEARER CONTEXT ACCEPT message, from the mobile to the MME entity.

When the network triggers default bearer restoration, the mobile receives an RRC *Paging* message from the evolved Node B (eNB) entity.

On receipt of this message, the mobile initializes the service request procedure described above.

Table 2.13 summarizes the ESM messages exchanged for the establishment, the modification and the release of the default bearer.

Source	Establishment of default bearer initialized by the mobile	Destination
UE	PDN CONNECTIVITY REQUEST	MME
MME	ACTIVATE DEFAULT EPS BEARER CONTEXT REQUEST or PDN CONNECTIVITY REJECT	UE
UE	ACTIVATE DEFAULT EPS BEARER CONTEXT ACCEPT	MME
Source	**Default bearer restoration**	**Destination**
MME	ACTIVATE DEFAULT EPS BEARER CONTEXT REQUEST	UE
UE	ACTIVATE DEFAULT EPS BEARER CONTEXT ACCEPT or ACTIVATE DEFAULT EPS BEARER CONTEXT REJECT	MME
Source	**Default bearer release initialized by the mobile**	**Destination**
UE	PDN DISCONNECT REQUEST	MME
MME	DEACTIVATE EPS BEARER CONTEXT REQUEST or PDN DISCONNECT REJECT	UE
UE	DEACTIVATE EPS BEARER CONTEXT ACCEPT	

Table 2.13. *ESM messages: default bearer case*

Dedicated bearer establishment may be initiated either by the mobile, or by the network.

When the network initiates dedicated bearer establishment, the mobile and the MME entity exchange the following messages:

– ESM ACTIVATE DEDICATED EPS BEARER CONTEXT REQUEST message transmitted by the MME entity;

– ESM ACTIVATE DEDICATED EPS BEARER CONTEXT ACCEPT message returned by the mobile.

When the mobile initiates dedicated bearer establishment, the mobile initiates the procedure by sending the ESM BEARER RESOURCE ALLOCATION REQUEST message to the MME entity.

Table 2.14 summarizes the ESM messages exchanged for the establishment, the modification and the release of the dedicated bearer.

Source	Dedicated bearer establishment initiated by the network	Destination
MME	ACTIVATE DEDICATED EPS BEARER CONTEXT REQUEST	UE
UE	ACTIVATE DEDICATED EPS BEARER CONTEXT ACCEPT or ACTIVATE DEDICATED EPS BEARER CONTEXT REJECT	MME
Source	**Dedicated bearer modification initiated by the network**	**Destination**
MME	MODIFY EPS BEARER CONTEXT REQUEST	UE
UE	MODIFY EPS BEARER CONTEXT ACCEPT or MODIFY EPS BEARER CONTEXT REJECT	MME
Source	**Dedicated bearer release initiated by the network**	**Destination**
MME	DEACTIVATE EPS BEARER CONTEXT REQUEST	UE
UE	DEACTIVATE EPS BEARER CONTEXT ACCEPT	MME
Source	**Dedicated bearer establishment initiated by the mobile**	**Destination**
UE	BEARER RESOURCE ALLOCATION REQUEST	MME
MME	ACTIVATE DEDICATED EPS BEARER CONTEXT REQUEST or BEARER RESOURCE ALLOCATION REJECT	UE
UE	ACTIVATE DEDICATED EPS BEARER CONTEXT ACCEPT	MME
Source	**Dedicated bearer modification initiated by the mobile**	**Destination**
UE	BEARER RESOURCE MODIFICATION REQUEST	MME
MME	MODIFY EPS BEARER CONTEXT REQUEST or BEARER RESOURCE MODIFICATION REJECT	UE
UE	MODIFY EPS BEARER CONTEXT ACCEPT	MME

Table 2.14. *ESM messages: dedicated bearer case*

2.2.2. Message structure

Table 2.15 presents the main information elements of the ESM PDN CONNECTIVITY REQUEST message.

Information elements	Description
PDN Type	IPv4, IPv6, IPv4 and IPv6
Access point name	Access Point Name (PGW entity) (optional)
Protocol configuration options	Configuration parameters requested by the mobile

Table 2.15. *Information elements of the ESM PDN CONNECTIVITY REQUEST message*

Table 2.16 presents the main information elements of the ESM PDN CONNECTIVITY REJECT message.

Information elements	Description
ESM cause	Bearer establishment reject cause

Table 2.16. *Information elements of the ESM REJECT PDN CONNECTIVITY message*

Table 2.17 presents the main information elements of the ESM PDN DISCONNECT REQUEST message.

Information elements	Description
Linked EPS bearer identity	Identity of the link between the default bearer and the dedicated bearer

Table 2.17. *Information elements of the ESM PDN DISCONNECT REQUEST message*

Table 2.18 presents the main information elements of the ESM PDN DISCONNECT REJECT message.

Information elements	Description
ESM cause	Default bearer release reject cause

Table 2.18. *Information elements of the ESM REJECT PDN DISCONNECT message*

Table 2.19 presents the main information elements of the ESM ACTIVATE DEFAULT EPS BEARER CONTEXT REQUEST message.

Information elements	Description
EPS QoS	Quality of service parameters
Access point name	Access Point Name (PGW entity)
PDN Address	IPv4 or IPv6 or IPV4 / IPV6
Protocol configuration options	Mobile configuration parameters

Table 2.19. *Information elements of the ESM ACTIVATE DEFAULT EPS BEARER CONTEXT REQUEST message*

ESM ACTIVATE DEFAULT EPS BEARER CONTEXT ACCEPT message contains no mandatory information elements.

Table 2.20 presents the main information elements of the ESM ACTIVATE DEFAULT EPS BEARER CONTEXT REJECT message.

Information elements	Description
ESM cause	Default bearer activation reject cause

Table 2.20. *Information elements of the ESM ACTIVATE DEFAULT EPS BEARER CONTEXT REJECT message*

Table 2.21 presents the main information elements of the ESM ACTIVATE DEDICATED EPS BEARER CONTEXT REQUEST message.

Information elements	Description
Linked EPS bearer identity	Identity of the link between the default bearer and the dedicated bearer
EPS QoS	Quality of service parameters
TFT	Flow identifier

Table 2.21. *Information elements of the ESM ACTIVATE DEDICATED EPS BEARER CONTEXT REQUEST message*

The ESM ACTIVATE DEDICATED EPS BEARER CONTEXT ACCEPT message contains no mandatory information elements.

Table 2.22 presents the main information elements of the ESM ACTIVATE DEDICATED EPS BEARER CONTEXT REJECT message.

Information elements	Description
ESM cause	Dedicated bearer activation reject cause

Table 2.22. *Information elements of the ESM ACTIVATE DEDICATED EPS BEARER CONTEXT REJECT message*

Table 2.23 presents the main information elements of the DEACTIVATE EPS BEARER CONTEXT ESM REQUEST message.

Information elements	Description
ESM cause	Dedicated bearer deactivation reject cause

Table 2.23. *Information elements of the ESM DEACTIVATE EPS BEARER CONTEXT REQUEST message*

ESM DEACTIVATE EPS BEARER CONTEXT ACCEPT message contains no mandatory information elements.

Table 2.24 presents the main information elements of the ESM MODIFY EPS BEARER CONTEXT REQUEST message.

Information elements	Description
New EPS QoS	Quality of service parameters
TFT	Flow identifier

Table 2.24. *Information elements of the ESM MODIFY EPS BEARER CONTEXT REQUEST message*

The ESM MODIFY EPS BEARER CONTEXT ACCEPT message contains no mandatory information elements.

Table 2.25 presents the main information elements of the ESM MODIFY EPS BEARER CONTEXT REJECT message.

Information elements	Description
ESM cause	Dedicated bearer modification reject cause

Table 2.25. *Information elements of the ESM MODIFY EPS BEARER CONTEXT REJECT message*

Table 2.26 presents the main information elements of the ESM BEARER RESOURCE ALLOCATION REQUEST message.

Information elements	Description
Linked EPS bearer identity	Identity of the link between the default bearer and the dedicated bearer
Traffic flow aggregate	Flow set identifiers
Required QoS traffic flow	Required quality of service

Table 2.26. *Information elements of the ESM BEARER RESOURCE ALLOCATION REQUEST message*

Table 2.27 presents the main information elements of the BEARER RESOURCE ALLOCATION REJECT ESM message.

Information elements	Description
ESM cause	Dedicated bearer establishment reject cause

Table 2.27. *Information elements of the ESM BEARER RESOURCE ALLOCATION REJECT message*

Table 2.28 presents the main information elements of the ESM BEARER RESOURCE MODIFICATION REQUEST message.

Information elements	Description
EPS bearer identity for packet filter	EPS bearer identity
Traffic flow aggregate	Flow set identifiers
Required QoS traffic flow	Required quality of service

Table 2.28. *Information elements of the ESM BEARER RESOURCE MODIFICATION REQUEST message*

Table 2.29 presents the main information elements of the ESM BEARER RESOURCE CHANGE REJECT message.

Information elements	Description
ESM cause	Dedicated bearer modification reject cause

Table 2.29. *Information elements of the ESM BEARER RESOURCE CHANGE REJECT message*

3

RRC Protocol

3.1. System information

The radio characteristics of the serving cell and of the neighboring cells, as well as the alerts are broadcast in the cell by system information.

The system information comprises a Master Information Block (MIB) message and a set of System Information Block (SIB) messages.

Table 3.1 provides the transport characteristics of the MIB message and of SIB messages.

SRB	RLC mode	Logical channel	Transport channel	Physical channel
Not applicable	TM	BCCH	BCH	PBCH
MasterInformationBlock				
Not applicable	TM	BCCH	DL-SCH	PDSCH
SystemInformationBlock				

Table 3.1. *Message transport related to the system information*

The System Information Radio Network Temporary Identifier (SI-RNTI) is used to retrieve the description of the resource allocated to the SIB messages in the Physical Downlink Shared Channel (PDSCH). This description is provided by the Physical Downlink Shared Channel (PDCCH).

3.1.1. *MIB message*

The information provided by the MIB message enables the mobile subsequently to analyze the PDCCH physical channel, initially to retrieve SIB messages. MIB messages contain the following parameters:

```
MasterInformationBlock
    dl-Bandwidth
    phich-Config
    systemFrameNumber
```

dl-Bandwidth: this parameter specifies the bandwidth of the radio channel for the downlink direction, expressed in numbers of Resource Block (RB) in the frequency domain.

phich-Config: this parameter defines the configuration of the Physical HARQ Indication Channel (PHICH).

systemFrameNumber: this parameter defines the eight most significant bits of the frame number. The two least significant bits are acquired implicitly when decoding the Physical Broadcast Channel (PBCH).

3.1.2. *SIB1 message*

The SIB1 message provides information that enables the mobile to assess whether it is allowed to access to the cell and that define other SIB scheduling information. The SIB1 message contains the following parameters:

```
SystemInformationBlockType1
    cellAccessRelatedInfo
    cellSelectionInfo
    p-Max
    freqBandIndicator
    schedulingInfoList
    tdd-Config
    si-WindowLength
    systemInfoValueTag
    ims-EmergencySupport
    cellSelectionInfo
```

cellAccessRelatedInfo: this parameter provides various information regarding the access to the cell, such as the identities of the networks, Mobile Country Code (MCC) and Mobile Network Code (MNC), which share the cell, the Tracking Area Code (TAC) and the E-UTRAN Cell Global Identifier (ECGI).

cellSelectionInfo: this parameter provides the lowest value of the Reference Signal Received Power (RSRP) to select the cell.

p-Max: this parameter provides the maximum power allowed for the uplink direction.

freqBandIndicator: this parameter defines the frequency band in which the cell functions. This information is complementary to that provided by the MIB *(dl-Bandwidth)* message because it determines the duplex spacing.

schedulingInfoList: this parameter provides information that defines the grouping in System Information (SI) of SIB2 to SIB13 messages and the frequency of each group. The SIB2 messages are always included in the first SI group.

tdd-Config: this parameter provides the configuration of the type-2 time frame and of the special subframe.

si-WindowLength: this parameter determines the frame and subframe numbers allocated to each SI.

systemInfoValueTag: this parameter provides an indication of a modification of SIB2 to SIB9 or SIB13 messages.

ims-EmergencySupport: this parameter was introduced in Release 9, and indicates whether the cell supports emergency calls to a phone that cannot attach to the Mobility Management Entity (MME).

cellSelectionInfo: this parameter, introduced in Release 9, provides the minimum value of the Reference Signal Received Quality (RSRQ) to select the cell.

The SIB1 message is transmitted in the frames whose System Frame Number (SFN) is a multiple of 8.

The SIB1 message is repeated in the frames whose SFN number is even.

3.1.3. *SIB2 message*

The SIB2 message provides information relating to radio resource configuration. The SIB2 message contains the following parameters:

```
SystemInformationBlockType2
    ac-BarringInfo
    radioResourceConfigCommon
    ue-TimersAndConstants
    freqInfo
    mbsfn-SubframeConfigList
    timeAlignmentTimerCommon
    ssac-BarringForMMTEL-Voice
    ssac-BarringForMMTEL-Video
    ac-BarringForCSFB
```

ac-BarringInfo: this parameter indicates whether the cell accepts emergency calls and provides values for timer calculation when establishing connections relating to signaling and traffic data.

radioResourceConfigCommon: this parameter provides configuration information for various physical signals and physical channels.

ue-TimersAndConstants: this parameter provides the value of various timers:

– the T300 timer relates to the procedure for establishing the Radio Resource Control (RRC) connection and indicates the latency of the mobile response to the RRC *ConnectionRequest* message;

– the T301 timer starts after the transmission of the RRC *ConnectionRe-EstablishmentRequest* message. The mobile switches to the idle state if it has not received a response when the timer expires;

– the T310 timer starts if the mobile has received N310 desynchronization indications. The mobile switches to the idle state if it has not received N311 synchronization indications;

– the T311 timer starts during the RRC connection restoration procedure. The mobile switches to the idle state if it does not find a cell to connect.

freqInfo: this optional parameter provides the frequency band used for the uplink direction. If this parameter is not provided, the bandwidth for the uplink direction is deducted from the *freqBandIndicator* of the SIB1 message.

MBSFN-SubframeConfigList: this parameter, introduced in Release 9, provides a list of frames reserved for Multimedia Broadcast Multicast Service (MBMS) sessions.

timeAlignmentTimerCommon: this parameter provides the timer value during which the mobile sees its time advance as valid.

SSAC-BarringForMMTEL-Voice: this parameter, introduced in Release 9, provides values for timer calculation when establishing a telephone communication.

SSAC-BarringForMMTEL-Video: this parameter, introduced in Release 9, provides values for timer calculation when establishing a videophone communication.

ac-BarringForCSFB: this parameter, introduced in Release 10, provides values for timer calculation when establishing the Circuit-Switched FallBack (CSFB) procedure.

3.1.4. *SIB3 message*

The SIB3 message provides information that enables the mobile to select a new cell. The SIB3 message contains the following parameters:

```
SystemInformationBlockType3
   cellReselectionInfoCommon
   cellReselectionServingFreqInfo
   intraFreqCellReselectionInfo
```

cellReselectionInfoCommon: this parameter provides the hysteresis value to use for the selection of a new cell.

cellReselectionServingFreqInfo: this parameter provides the conditions for the selection of a new 4G cell with different frequency (inter-frequency) from that of the serving cell or a new 2G/3G cell (inter-Radio Access Technology (RAT)).

intraFreqCellReselectionInfo: this parameter provides the conditions for the selection of a new 4G cell with the same frequency (intra-frequency) as the serving cell. Release 9 introduced new elements.

3.1.5. *SIB4 message*

The SIB4 message provides information relating to the neighboring 4G cells with the same frequency (intra-frequency) as the serving cell. The SIB4 message contains the following parameters:

```
SystemInformationBlockType4
   intraFreqNeighCellList
   intraFreqBlackCellList
   csg-PhysCellIdRange
```

The three parameters are optional; the mobile has the ability to select a new intra-frequency cell without this information.

intraFreqNeighCellList: this parameter provides a list of the neighboring cells identified by their Physical-layer Cell Identity (PCI).

intraFreqBlackCellList: this parameter provides a list of the prohibited neighboring cells.

csg-PhysCellIdRange: this parameter provides a list of the reserved neighboring cells, such as Home evolved Node B (HeNB) cells.

3.1.6. *SIB5 message*

The SIB5 message provides information relating to the neighboring 4G cells with different frequency (inter-frequency) from that of the serving cell:

```
SystemInformationBlockType5
   interFreqCarrierFreqList
```

3.1.7. *SIB6 message*

The SIB6 message provides information relating to the neighboring cells of the 3G Universal Terrestrial Radio Access Network (UTRAN). The SIB6 message contains the following parameters:

```
SystemInformationBlockType6
   carrierFreqListUTRA-FDD
   carrierFreqListUTRA-TDD
   t-ReselectionUTRA
   t-ReselectionUTRA-SF
```

carrierFreqListUTRA-FDD: this parameter provides information about the neighboring cells functioning according to the Frequency Division Duplex (FDD) mode.

carrierFreqListUTRA-TDD: this parameter provides information on the neighboring cells functioning according to the Time Division Duplex (TDD) mode.

t-ReselectionUTRA: this parameter provides a timer value for 3G UTRAN cell selection.

t-ReselectionUTRA-SF: this parameter provides the multiplicative factor applied to the parameter *t-ReselectionUTRA.*

3.1.8. *SIB7 message*

The SIB7 message provides information relating to the neighboring cells of the 2G GSM/EDGE Radio Access Network (GERAN). The SIB7 message contains the following parameters:

```
SystemInformationBlockType7
   t-ReselectionGERAN
   t-ReselectionGERAN-SF
   carrierFreqsInfoList
```

t-ReselectionGERAN: this parameter provides a timer value for selection of the 2G GERAN cell.

t-ReselectionGERAN-SF: this parameter provides the multiplicative factor applied to the parameter *t-ReselectionGERAN.*

carrierFreqsInfoList: this parameter provides information relating to the neighboring 2G GERAN cells.

3.1.9. *SIB8 message*

The SIB8 message provides information relating to the neighboring cells of the 3G Code Division Multiple Access (CDMA2000). The SIB8 message contains the following parameters:

```
SystemInformationBlockType8
    systemTimeInfo
    searchWindowSize
    parametersHRPD
    parameters1XRTT
```

systemTimeInfo: this parameter indicates whether the radio access networks 3G CDMA2000 and 4G E-UTRAN are synchronized.

searchWindowSize: this parameter provides support to the mobile to search for 3G CDMA2000 radio access network pilots.

parametersHRPD: this parameter provides information relating to the neighboring cells of the 3G CDMA2000 access network operating in High Rate Packet Data (HRPD) mode.

parameters1XRTT: this parameter provides information relating to the neighboring cells of the 3G CDMA2000 access network operating in Radio Transmission Technology (1xRTT) mode.

3.1.10. *SIB9 message*

The SIB9 message provides the name of the home HeNB cell:

```
SystemInformationBlockType9
    hnb-Name
```

3.1.11. *SIB10 message*

For the Japanese version, the SIB10 message provides succinct information relating to Earthquake and Tsunami Warning System (ETWS) alerts. The SIB10 message contains the following parameters:

```
SystemInformationBlockType10
    messageIdentifier
    serialNumber
    warningType
```

messageIdentifier: this parameter provides the notification type.

serialNumber: this parameter provides a record of the modifications of various notifications.

warningType: this parameter indicates whether the notification relates to an earthquake only, a tsunami only, or an earthquake and a tsunami.

3.1.12. *SIB11 message*

The SIB11 message provides more detailed information relating to ETWS alerts. The SIB11 message contains the following parameters:

```
SystemInformationBlockType11
   messageIdentifier
   serialNumber
   warningMessageSegmentType
   warningMessageSegmentNumber
   warningMessageSegment
   dataCodingScheme
```

messageIdentifier: this parameter provides the notification type.

serialNumber: this parameter provides a record of the modifications of various notifications.

warningMessageSegmentType: this parameter indicates whether the *warningMessageSegment* parameter is the last segment constituting the message.

warningMessageSegmentNumber: this parameter provides the segment number.

warningMessageSegment: this parameter contains the alert message segment.

dataCodingScheme: this parameter defines the syntax used for the alert message.

3.1.13. *SIB12 message*

The SIB12 message introduced in Release 9 provides, for the North American version, the Commercial Mobile Alert System (CMAS) information. The SIB 12 message contains the same parameters as the SIB11 message:

```
SystemInformationBlockType12
    messageIdentifier
    serialNumber
    warningMessageSegmentType
    warningMessageSegmentNumber
    warningMessageSegment
    dataCodingScheme
```

3.1.14. *SIB13 message*

The SIB13 message, introduced in Release 9, provides information relating to the MBMS over Single Frequency Network (MBSFN) and contains the following parameters:

```
SystemInformationBlockType13
    mbsfn-AreaInfoList
    notificationConfig
```

MBSFN-AreaInfoList: this parameter provides the configuration of the Multicast Control Channel (MCCH).

notificationConfig: this parameter provides the notification configuration of the MCCH logical channel.

3.2. Connection control

3.2.1. *Paging*

The paging procedure concerns the following operations:

– indication of incoming data to the mobile in the RRC-IDLE state;

– indication of a telephone call from the network in Circuit-Switched (CS) mode to the mobile in the RRC-IDLE state;

– indication of a modification of system information to mobiles in the RRC-IDLE state or the RRC-CONNECTED state;

– notification of Earthquake and Tsunami Warning System (ETWS);

– notification of Commercial Mobile Alerting System (CMAS).

Table 3.2 provides the transport characteristics of the paging message.

SRB	RLC mode	Logical channel	Transport channel	Physical channel
Not applicable	TM	PCCH	PCH	PDSCH
Paging				

Table 3.2. *Message transport relating to the paging*

The paging message is transmitted by the evolved Node B (eNB) entity and contains the following information:

```
Paging
   pagingRecordList
   systemInfoModification
   etws-Indication
   cmas-Indication
```

pagingRecordList: this information element indicates the identity of the mobile which is the recipient of the notification, Shortened Temporary Mobile Subscriber Identity (S-TMSI) or International Mobile Subscriber Identity (IMSI), and the origin of the notification, with the network in CS mode or Packet-Switched (PS) mode.

systemInfoModification: this information element indicates whether the mobile has to acquire the new configuration of the System Information Block (SIB) message.

etws-Note: this information element indicates whether the mobile has to acquire the new configuration of the SIB10 and SIB11 message.

cmas-Note: this information element indicates whether the mobile has to acquire the new configuration of the SIB12 message.

3.2.2. *Connection establishment*

The connection establishment follows after the random access procedure and allows the transition from the RRC-IDLE state to the RRC CONNECTED state and the establishment of the Signaling Radio Bearer 1 (SRB1).

Table 3.3 provides the transport characteristics of the messages relating to the connection establishment.

SRB	RLC mode	Logical channel	Transport channel	Physical channel
SRB0	TM	CCCH	UL-SCH	PUSCH
ConnectionRequest				
SRB0	TM	CCCH	DL-SCH	PDSCH
ConnectionSetup *ConnectionReject*				
SRB1	AM	DCCH	UL-SCH	PUSCH
ConnectionSetupComplete				

Table 3.3. *Message transport relating to the connection establishment*

The *ConnectionRequest* message is transmitted by the mobile to initiate the connection establishment process and contains the following information elements:

```
RRC ConnectionRequest
   ue-Identity
   establishmentCause
```

ue-Identity: this information element indicates the mobile identity, the S-TMSI identity if the mobile is attached or a random value otherwise.

establishmentCause: this information element indicates the cause of the establishment of the connection.

The *ConnectionSetup* message is the response of the eNB entity to accept the request for the establishment of the connection and contains the following information elements:

```
RRCConnectionSetup
   rrc-TransactionIdentifier
   radioResourceConfigDedicated
```

rrc-TransactionIdentifier: this information element provides the transaction identifier.

radioResourceConfigDedicated: this information element provides the SRB1 bearer characteristics.

The *ConnectionReject* message is the response of the eNB entity to reject the request for the establishment of the connection and contains the following information elements:

```
RRCConnectionReject
   waitTime
   extendedWaitTime
```

waitTime: this information element indicates that the wait time is between 1 and 16 seconds.

extendedWaitTime: this information element, introduced in Release 10, extends the wait time to 1800 seconds.

The *ConnectionSetupComplete* message is a response of the mobile to the *ConnectionSetup* message to confirm the establishment of the connection and contains the following information elements:

```
RRCConnectionSetupComplete
   rrc-TransactionIdentifier
   selectedPLMN-Identity
   registeredMME
   dedicatedInfoNAS
```

selectedPLMN-Identity: this information element defines the identity of the network, Mobile Country Code (MCC) and Mobile Network Code (MNC), to which the mobile wants to connect, if the cell is shared between multiple networks.

registeredMME: this information element indicates the Globally Unique MME Identity (GUMMEI) which has attached the mobile.

dedicatedInfoNAS: this information element indicates that the RRC message carries a Non-Access Stratum (NAS) message.

3.2.3. *Security activation*

The security activation process starts when the eNB entity receives a setup request from the MME entity for the establishment of the mobile context.

The security activation procedure of the radio interface allows the implementation of the following operations:

– Data Radio Bearer (DRB) encryption;

– encryption and integrity control of the SRB1 and SRB2 bearers.

Table 3.4 provides the transport characteristics of the messages relating to the security activation.

<table>
<tr><td>SRB</td><td>RLC mode</td><td>Logical channel</td><td>Transport channel</td><td>Physical channel</td></tr>
<tr><td>SRB1</td><td>AM</td><td>DCCH</td><td>DL-SCH</td><td>PDSCH</td></tr>
<tr><td colspan="5">SecurityModeCommand</td></tr>
<tr><td>SRB1</td><td>AM</td><td>DCCH</td><td>UL-SCH</td><td>PUSCH</td></tr>
<tr><td colspan="5">SecurityModeComplete
SecurityModeFailure</td></tr>
</table>

Table 3.4. *Message transport relating to the security activation*

The *SecurityModeCommand* message is transmitted by the eNB entity to define the algorithms used to encrypt the SRB and DRB bearers and for the integrity control of SRB bearer.

The *SecurityModeComplete* message is transmitted by the mobile to confirm security activation.

The *SecurityModeFailure* message is transmitted by the mobile to indicate a security activation failure.

3.2.4. *Connection reconfiguration*

The connection reconfiguration procedure covers the following operations:

– establishment, modification and release of DRB bearers;

– establishment, modification and release of radio measurements;

– establishment, modification and release of Secondary Cells (SCell);

– handover triggering.

Table 3.5 provides the transport characteristics of the messages relating to the connection reconfiguration.

SRB	RLC mode	Logical channel	Transport channel	Physical channel
SRB1	AM	DCCH	DL-SCH	PDSCH
		ConnectionReconfiguration		
SRB1	AM	DCCH	UL-SCH	PUSCH
		ConnectionReconfigurationComplete		

Table 3.5. *Message transport relating to the connection reconfiguration*

The *ConnectionReconfiguration* message is transmitted by the eNB entity and contains the following information elements:

```
RRCConnectionReconfiguration
   measConfig
   mobilityControlInfo
   dedicatedInfoNASList
   radioResourceConfigDedicated
   securityConfigHO
   sCellToReleaseList
   sCellToAddModList
```

measConfig: this information element defines the measurements to be carried out on the neighboring 4G cells (intra-frequency and inter-frequency) and on the neighboring 2G and 3G (inter-RAT) cells.

mobilityControlInfo: this information element provides the characteristics of the target cell in the case of handover.

dedicatedInfoNAS: this information element is used to transfer a NAS message.

radioResourceConfigDedicated: this information element is used for the establishment, the modification and the release of the radio bearer and for the change of the configuration of the Medium Access Control (MAC) and of the dedicated physical channels.

securityConfigHO: this information element allows the renewal of integrity control keys and encryption keys during the handover.

sCellToReleaseList: this information element, introduced in Release 10, indicates the list of Secondary Cells (SCell) to remove.

sCellToAddModList: this information element, introduced in Release 10, indicates the list of Secondary Cells (SCell) to add.

The *ConnectionReconfigurationComplete* message is transmitted by the mobile to confirm connection reconfiguration.

3.2.5. *Connection re-establishment*

Connection re-establishment allows the resumption of the SRB1 bearer operation, of the reactivation of the security and of the primary PCell configuration only.

Table 3.6 provides the transport characteristics of the message relating to the connection re-establishment.

<table>
<tr><th>SRB</th><th>RLC mode</th><th>Logical channel</th><th>Transport channel</th><th>Physical channel</th></tr>
<tr><td>SRB0</td><td>TM</td><td>CCCH</td><td>UL-SCH</td><td>PUSCH</td></tr>
<tr><td colspan="5">ConnectionRe-establishmentRequest</td></tr>
<tr><td>SRB0</td><td>TM</td><td>CCCH</td><td>DL-SCH</td><td>PDSCH</td></tr>
<tr><td colspan="5">ConnectionRe-establishment
ConnectionRe-establishmentReject</td></tr>
<tr><td>SRB1</td><td>AM</td><td>DCCH</td><td>UL-SCH</td><td>PUSCH</td></tr>
<tr><td colspan="5">ConnectionRe-establishmentComplete</td></tr>
</table>

Table 3.6. *Message transport relating to the connection re-establishment*

The *ConnectionRe-establishmentRequest* message is transmitted by the mobile and contains the following information elements:

```
RRCConnectionReestablishmentRequest
   ue-Identity
   reestablishmentCause
```

ue-Identity: this information element contains the Cell Radio Network Temporary Identity (C-RNTI) of the mobile and the Physical-layer Cell Identity (PCI) to which the mobile was connected before failure.

re-establishmentCause: this information element indicates the cause (handover, reconfiguration) of the connection re-establishment triggering.

The *ConnectionRe-establishment* message is transmitted by the eNB entity and contains the following information elements:

```
RRCConnectionReestablishment
   radioResourceConfigDedicated
   nextHopChainingCount
```

radioResourceConfigDedicated: this information element provides the characteristics of the radio bearers to re-establish.

nextHopChainingCount: this information element provides the parameter that allows the renewal of encryption key and integrity control.

The *ConnectionRe-establishmentReject* message is transmitted by the eNB entity to indicate the rejection of the connection re-establishment request.

The *ConnectionRe-establishmentComplete* message is transmitted by the mobile to confirm connection re-establishment.

3.2.6. *Connection release*

The connection release procedure allows the release of radio resources.

Table 3.7 provides the transport characteristics of the message relating to the connection release.

SRB	RLC mode	Logical channel	Transport channel	Physical channel
SRB1	AM	DCCH	DL-SCH	PDSCH
ConnectionRelease				

Table 3.7. *Message transport relating to the connection release*

The *ConnectionRelease* message is transmitted by the eNB entity and contains the following information elements:

```
RRCConnectionRelease
   releaseCause
   redirectedCarrierInfo
   idleModeMobilityControlInfo
```

releaseCause: this information element indicates the cause of connection release.

redirectedCarrierInfo: this information element allows the redirection of the mobile to another 4G or 2G/3G radio channel.

idleModeMobilityControlInfo: this information element provides the priority levels for the selection of new cells.

3.3. Measurements

3.3.1. *Introduction*

Measurements carried out on the serving cell and on the neighboring cells are used for the selection of the cell and for the handover.

The intra-frequency measurement, essential for mobility, is configured when connecting.

The inter-frequency measurement and inter-RAT (Radio Access Technology) can also be configured when connecting.

As the mobile does not generally have several radio receivers, the inter-frequency and inter-RAT measurement should be performed at intervals arranged in the frame.

The configuration of measurements to be performed by the mobile is triggered by the eNB entity in RRC *ConnectionSetup, ConnectionReconfiguration, ConnectionRe-establishment* messages.

The measurement configurations to perform define the following parameters:

– object that identifies the radio channel;

– event triggering the measurement report;

– combination of objects and events;

– measurement filtering parameter;

– periodicity of measurements.

3.3.2. *Objects*

The measurements made on the radio channels of the neighboring cells are of different types:

– intra-frequency: the frequency of the neighboring cell is identical to that of the serving cell;

– inter-frequency: the frequency of the neighboring cell is different from that of the serving cell;

– inter-RAT Universal Terrestrial Radio Access Network (UTRAN);

– inter-RAT GSM / EDGE Radio Access Network (GERAN);

– inter-RAT Code Division Multiple Access (CDMA2000).

The 4G neighboring cells are specified by their Physical-layer Cell Identity (PCI) and possibly E-UTRAN Cell Global Identity (ECGI).

The UTRAN neighboring cells are specified by their Base Station Identity Code (BSIC), by the scrambling code and optionally by their Cell Global Identity (CGI).

The GERAN neighboring cells are specified by their BSIC code and possibly by their CGI identity.

The CDMA2000 neighboring cells are specified by their scrambling Pseudo-random Noise (PN) and possibly by their CGI identity.

The measurements on the neighboring cells relate to the following values:

– for 4G cells, Reference Signal Received Power (RSRP) and Reference Signal Received Quality (RSRQ) measured on the reference signal;

– for 3G cells, Received Signal Code Power (RSCP) and quality level measured on the Common Pilot Channel (CPICH);

– for 2G cells, Received Signal Strength Indication (RSSI).

3.3.3. *Events*

3.3.3.1. *Event A1*

Event A1 occurs when the measured value of the signal received from the serving cell becomes higher than a threshold:

$$Ms - Hys > Thresh$$

Ms is the measured value of the signal received on the serving cell.

Hys is the hysteresis parameter for this event.

Thresh is the threshold for this event.

Event A1 is stopped when the measured value of the signal received from the serving cell becomes less than a threshold:

$$Ms + Hys < Thresh$$

3.3.3.2. *Event A2*

Event A2 occurs when the measured value of the signal received from the serving cell becomes less than a threshold:

$$Ms + Hys < Thresh$$

Event A2 is stopped when the measured value of the signal received from the serving cell becomes greater than a threshold:

$$Ms - Hys > Thresh$$

3.3.3.3. *Event A3*

Event A3 occurs when the measured value of the signal received from the neighboring cell is greater than that of the serving cell:

$$Mn + Ofn + Ocn - Hys > Mp + Ofp + Ocp + Off$$

Mn is the measured value of the signal received from the neighboring cell.

Ofn is the specific offset linked to the frequency of the neighboring cell.

Ocn is the specific offset linked to the neighboring cell.

Mp is the measured value of the signal received from the serving cell.

Ofp is the specific offset linked to the frequency of the serving cell.

OCP is the specific offset linked to the serving cell.

Off is the specific offset to this event.

For Release 10, *Mp* is the measured value of the level received on the primary channel PCell of the serving cell.

Event A3 is stopped when the measured value of the signal received from the neighboring cell is less than that of the serving cell:

$$Mn + Ofn + Ocn + Hys < Mp + Ofp + Ocp + Off$$

3.3.3.4. *Event A4*

Event A4 occurs when the measured value of the signal received from the neighboring cell becomes greater than a threshold:

$$Mn + Ofn + Ocn - Hys > Thresh$$

Event A4 is stopped when the measured value of the signal received from the neighboring cell becomes lower than a threshold:

$$Mn + Ofn + Ocn - Hys > Thresh$$

3.3.3.5. *Event A5*

Event A5 occurs when the measured value of the signal received from the serving cell becomes less than a first threshold and the measured value of the signal received from the neighboring cell becomes greater than a second threshold:

$$Mp + Hys < Thresh1$$

$$Mn + Ofn + Ocn - Hys > Thresh2$$

Event A5 is stopped when the measured value of the signal received from the serving cell becomes greater than a first threshold and the measured value of the signal received from the neighboring cell becomes less than a second threshold:

$$Mp - Hys > Thresh1$$

$$Mn + Ofn + Ocn + Hys < Thresh2$$

3.3.3.6. *Event A6*

Event A6, introduced in Release 10, occurs when the measured value of the signal received from the neighboring cell becomes greater than that of the signal received from the serving cell, for the secondary channel SCell:

$$Mn + Ocn - Hys > Ms + Ocs + Off$$

Ms is the measured value of the signal received from the secondary channel SCell of the serving cell.

OCS is the specific offset linked to the secondary channel SCell of the serving cell.

Event A6 is stopped when measured value of the signal received from the neighboring cell becomes less than that of the signal received from the serving cell, for the secondary channel SCell:

$$Mn + Ocn + Hys < Ms + Ocs + Off$$

3.3.3.7. *Event B1*

Event B1 occurs when the measurement of the signal received from the neighboring 2G or 3G cell becomes greater than a threshold:

$$Mn + Ofn - Hys > Thresh$$

Event B1 is stopped when the measured value of the signal received from the neighboring 2G or 3G cell becomes less than a threshold:

$$Mn + Ofn + Hys < Thresh$$

3.3.3.8. *Event B2*

Event B2 occurs when the measured value of the signal received from the serving cell becomes less than a first threshold and the measured value of the signal received from the neighboring 2G or 3G cell becomes greater than a second threshold:

$$Mp + Hys < Thresh1$$

$$Mn + Ofn - Hys > Thresh2$$

The event B2 is stopped when the measured value of the signal received from the serving cell becomes greater than a first threshold and the measurement of the signal received from the neighboring 2G or 3G cell becomes less than a second threshold:

$$Mp - Hys > Thresh1$$

$$Mn + Ofn + Hys < Thresh2$$

3.3.4. *Measurement filtering*

The measurement filtering consists in weighting the value communicated by the physical layer.

The filtering of each type of measurement is obtained by applying the following relationship:

$$F_n = (1-a) \cdot F_{n-1} + a \cdot M_n$$

F_n is the updated filtered measurement.

F_{n-1} is the previous measurement, F_0 corresponding to the first measurement M_1 received from the physical layer.

$a = 1/2^{(k/4)}$, where k is the filtering coefficient.

M_n is the last measurement received from the physical layer.

3.3.5. *Measurement report*

The measurement reports can be triggered by an event or returned periodically or periodically after an event.

Table 3.8 provides the transport characteristics of the message relating to the measurement report.

SRB	RLC mode	Logical channel	Transport channel	Physical channel
SRB1	AM	DCCH	UL-SCH	PUSCH
MeasurementReport				

Table 3.8. *Message transport relating to the measurement report*

3.4. Broadcast control

3.4.1. *Configuration of frames and subframes*

In Release 9, the System Information Block 2 (SIB2) introduced an information element *MBSFN-SubframeConfig* defining subframes assigned to the Multimedia Broadcast Multicast Service (MBMS).

```
MBSFN-SubframeConfig
    radioframeAllocationPeriod
    radioframeAllocationOffset
    subframeAllocation
```

radioframeAllocationPeriod: this field contains the value of the frame allocation periodicity to the MBMS service (Figure 3.1).

radioframeAllocationOffset: this parameter contains the value of the offset of the frame allocated to the MBMS service (Figure 3.1):

SFN mod *radioFrameAllocationPeriod* = *radioFrameAllocationOffset*

subframeAllocation: this parameter defines the subframes (among the subframes 1, 2, 3, 6, 7, 8) assigned to the MBMS service (Figure 3.1).

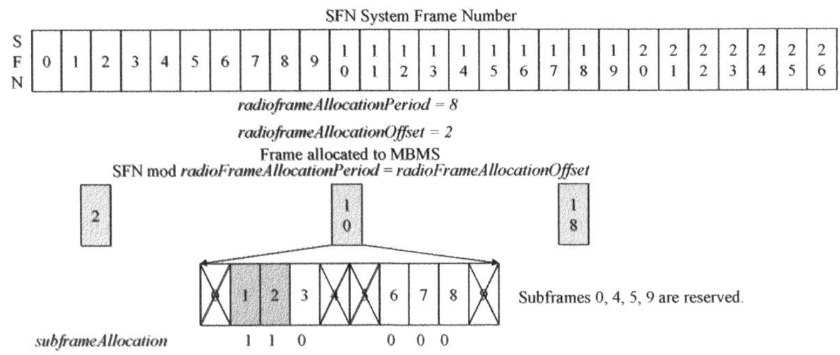

Figure 3.1. *Allocation of frames and subframes to the MBMS service*

3.4.2. *MCCH logical channel scheduling*

The SIB13 system information, introduced in Release 9, provides information relating to the Multicast Control Channel (MCCH) scheduling from the *mcch-Config* information element:

```
MCCH Config
    MCCH-RepetitionPeriod
    MCCH-Offset
    MCCH-ModificationPeriod
    sf-AllocInfo
    signallingMCS
```

mcch-RepetitionPeriod: this field defines the repetition period, by number of frames, of the MCCH logical channel (Figure 3.2).

mcch-Offset: this field sets the offset value of the frame where the MCCH logical channel is located:

SFN mod *mcch-RepetitionPeriod* = *mcch-Offset*

mcch-ModificationPeriod: this field defines the period during which there is no modification of the MCCH logical channel (Figure 3.2). The limits of the period without modification will satisfy the following relationship:

SFN mod *mcch-ModificationPeriod* = *0*

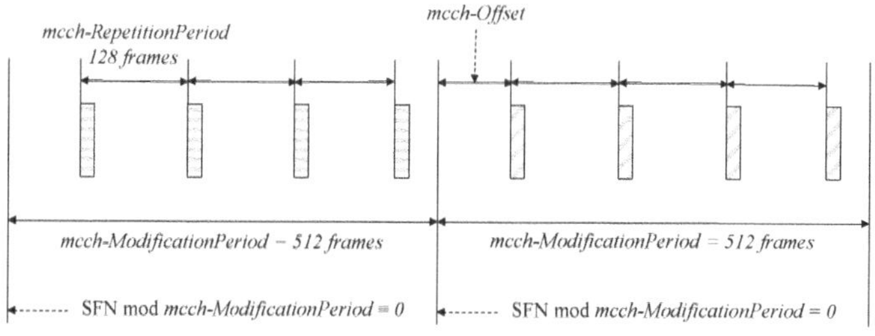

Figure 3.2. *MCCH logical channel scheduling*

The modification is signaled to the mobile by Downlink Control Information (DCI) in the 1C format, carried by the Physical Downlink Control Channel (PDCCH).

sf-AllocInfo: this field defines the subframe where the MCCH logical channel is located.

signallingMCS: this field defines the Modulation and Coding Scheme (MCS) applied to the data relating to the MCCH logical channel.

3.4.3. *MTCH logical channel scheduling*

The MCCH physical channel contains the Radio Resource Control (RRC) message that defines the scheduling of the Multicast Traffic Channel (MTCH).

Table 3.9 provides the transport characteristics of the message relating to the MCCH logical channel scheduling.

SRB	RLC mode	Logical channel	Transport channel	Physical channel
not applicable	UM	MCCH	MCH	PMCH
MBSFNAreaConfiguration				

Table 3.9. *Message transport relating to the MTCH logical channel scheduling*

The *MBSFNAreaConfiguration* message is transmitted by the eNB entity and contains the following information:

```
MBSFNAreaConfiguration
   commonSF-Alloc
   commonSF-AllocPeriod
   PMCH-infolist
```

commonSF-Alloc: this field defines the frames and the subframes allocated to each MBMS Single Frequency Network (MBSFN) area (Figure 3.3).

commonSF-AllocPeriod: this field defines the periodicity of the allocation pattern of subframes common to all MTCH logical channels of the MBSFN area (Figure 3.3).

PMCH-infolist: this field defines the multiplexing of several MTCH logical channels in each Physical Multicast Channel (PMCH) allocated to an MBSFN area and the modulation and coding scheme applied to the PMCH physical channel.

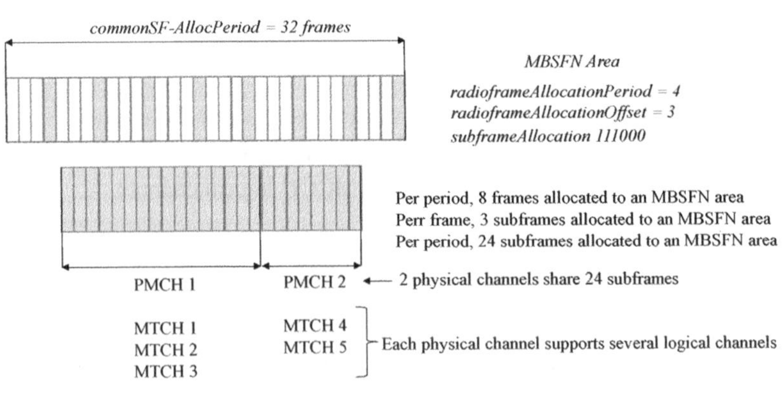

Figure 3.3. *MTCH logical channel scheduling*

3.4.4. *Counting*

The counting messages were introduced in Release 10.

The counting procedure helps the Multi-cell/Multicast Coordination Entity (MCE) choose the mode of transmission of the MTCH physical channel, either broadcast mode or unicast mode:

– the unicast mode is most effective if the number of mobiles subscribed in a program is low. The unicast mode allows the use of the Multiple Input Multiple Output (MIMO) transmission mode and the optimization of the modulation and coding scheme;

– the broadcast mode is more effective if the number of mobiles subscribed in a program is high. The broadcast mode only transmits a single logical channel regardless of the number of mobiles subscribed in the program.

Table 3.10 provides the transport characteristics of the messages relating to the counting.

SRB	RLC mode	Logical channel	Transport channel	Physical channel
Not applicable	UM	MCCH	MCH	PMCH
MBMSCountingRequest				
SRB1	AM	DCCH	UL-SCH	PUSCH
MBMSCountingResponse				

Table 3.10. *Message transport relating to the counting*

The *MBMSCountingRequest* message is transmitted by the eNB entity and contains a list of programs.

The *MBMSCountingResponse* message is transmitted by each mobile subscribed to one of the programs.

4

Data Link Layer

4.1. PDCP protocol

The Packet Data Convergence Protocol (PDCP) is used for Radio Resource Control (RRC), relating to dedicated control data, and for traffic data relating to Internet Protocol (IP) packet.

The PDCP protocol performs the following functions:

– compression of traffic data headers using the Robust Header Compression (ROHC) mechanism;

– security of traffic data (confidentiality) and of RRC (integrity and confidentiality);

– delivery in sequence of RRC messages and IP packets;

– recovery of PDCP frames lost during the handover.

Several PDCP instances can be activated simultaneously:

– two instances for Signaling Radio Bearers SRB1 and SRB2;

– SRB1 bearer is used for the transmission of an RRC message that can support a Non-Access Stratum (NAS) message;

– SRB2 bearer is used for the transmission of a NAS message only;

– one instance, for each Data Radio Bearer (DRB) relating to traffic data.

4.1.1. *Procedures*

4.1.1.1. *Security architecture*

Figure 4.1 describes the implementation of security keys used for traffic data and RRC signaling.

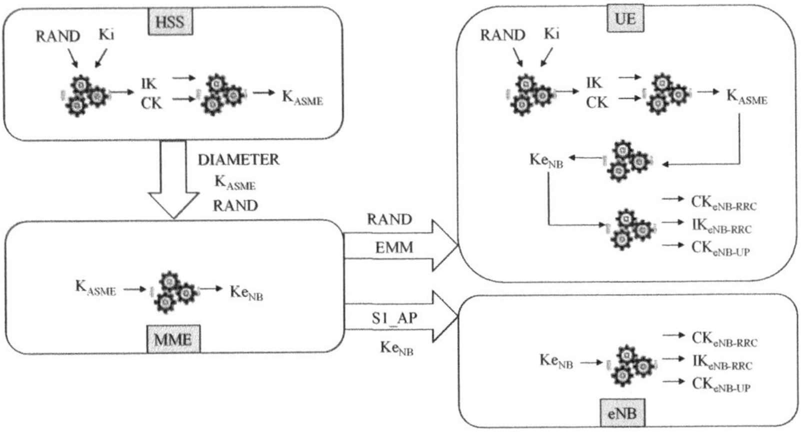

Figure 4.1. *Key generation*

During the procedure of attachment and development of the Authentication and Key Agreement (AKA) mechanism, the Home Subscriber Server (HSS) entity retrieves the Ki key of the mobile, draws a Random Number (RAND) and calculates the Integrity Key (IK) and Cipher Key (CK).

The Ki Key is a secret key, shared between the HSS entity and the mobile, created during subscription.

The HSS entity calculates the K_{ASME} key from the IK and CK keys and transmits the K_{ASME} key and the random number to the Mobility Management Entity (MME) in a DIAMETER message.

The MME entity calculates the Ke_{NB} key from the K_{ASME} key, and transmits it to the evolved Node B (eNB) entity in an S1-AP message.

The eNB entity calculates the following keys from the Ke_{NB} key:

– $CK_{eNB-RRC}$ used for the encryption of RRC messages;

– $IK_{eNB-RRC}$ used for the integrity control of RRC messages;

– $CK_{eNB\text{-}UP}$ is used for the encryption of traffic data.

The MME entity transmits the random number to the mobile in an EPS Mobility Management (EMM) message.

From the random number and the Ki key stored in the Universal Subscriber Identity Module (USIM) of its Universal Integrated Circuit Card (UICC), the mobile meanwhile performs the same operations as the HSS, MME and eNB entities to calculate the security keys used for RRC messages and traffic data.

4.1.1.2. *Header compression*

The header compression is based on the ROHC mechanism for which multiple algorithms have been defined by the Request For Comments (RFC) specifications of the Internet Engineering Task Force (IETF) standards body (Table 4.1).

Profile identifier	Compressed headers	References
0×0000	uncompressed	RFC 4995
0×0001	RTP/UDP/IP	RFC 3095, RFC 4815
0×0002	UDP/IP	RFC 3095, RFC 4815
0×0003	ESP/IP	RFC 3095, RFC 4815
0×0004	IP	RFC 3843, RFC 4815
0×0006	TCP/IP	RFC 4996
0×0101	RTP/UDP/IP	RFC 5225
0×0102	UDP/IP	RFC 5225
0×0103	ESP/IP	RFC 5225
0×0104	IP	RFC 5225

Table 4.1. *Compression protocols*

The header compression is based on the observation that in a session, a number of fields are invariant, such as IP addresses or port numbers.

The header compression is particularly effective when the size of the payload is relatively low, which is the case for voice (Figure 4.2).

The decompressor uses the PDCP Feedback control message to inform the compressor that the decompression is successful or that the synchronization between compression and decompression has been lost.

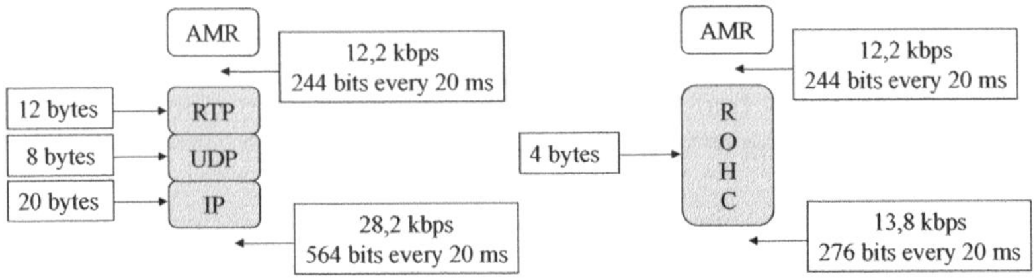

Figure 4.2. *Header compression*

4.1.1.3. *Loss of frames during handover*

The PDCP protocol allows recovery of frames lost during handover, only if the Radio Link Control (RLC) protocol uses the acknowledgement mode.

For the downlink direction, during the handover procedure, the source eNB entity sends the target eNB entity an X2-AP SN STATUS NUMBER message in which it provides, on the one hand, the PDCP frames transmitted to the mobile and unacknowledged with their Sequence Number (SN), and, on the other hand, the SN of non-received PDCP frames from the mobile for the uplink direction.

When the mobile connects to the target eNB entity, the recovery of lost frames is performed by the following exchanges:

– the mobile sends the target eNB entity the PDCP status report message that mentions the missing PDCP frames for the downlink direction;

– the target eNB entity sends the PDCP Status Report message to the mobile which mentions the missing PDCP frames for the uplink direction.

4.1.2. *Operations*

4.1.2.1. *Operations relating to the SRB bearer*

Figure 4.3 describes the operations relating to the SRB bearer.

At transmission, the RRC messages are subject to a numbering procedure which allows, at the receiver level, the detection of duplicate messages and the ordering.

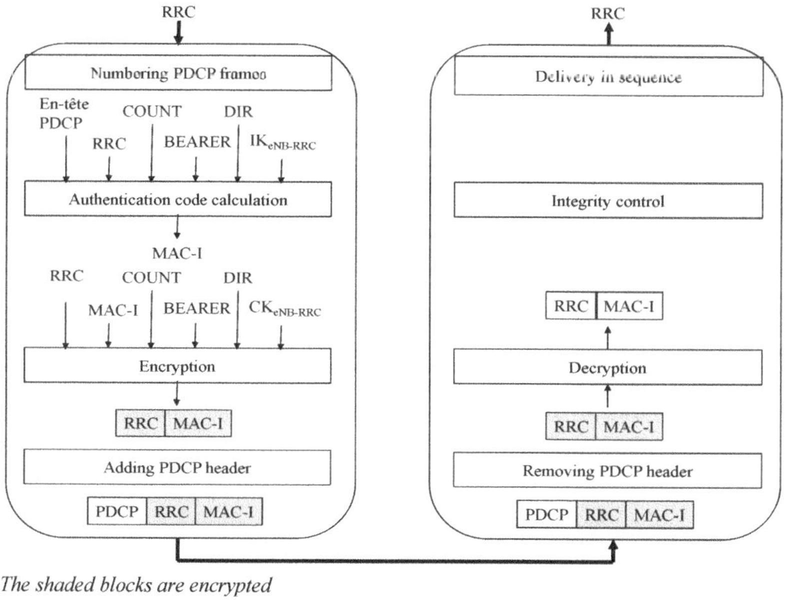

Figure 4.3. *PDCP operations relating to the SRB bearer*

The Message Authentication Code for Integrity (MAC-I) allows integrity control of the PDCP frame.

The calculation of the MAC-I message authentication code is done from the following:

– the PDCP frame that contains the PDCP header and RRC message;

– the COUNT parameter which aggregates the Hyper Frame Number (HFN) and the PDCP frame sequence number;

– the DIR parameter that indicates whether the data are transmitted in the downlink or uplink direction;

– the BEARER parameter which is provided by System Information 2 (SIB2);

– the $IK_{eNB\text{-}RRC}$ key.

The RRC message and the MAC-I authentication code are then encrypted using the $CK_{eNB\text{-}RRC}$ key, then encapsulated by the PDCP header to form the PDCP frame transmitted to the RLC layer.

On receiving the PDCP frames from the RLC layer, the following operations occur:

– the PDCP header is removed;

– the RRC message and the MAC-I authentication code are decrypted;

– the RRC message and the PDCP header are integrity controlled;

– the RRC messages are delivered in sequence.

4.1.2.2. *Operations relating to the DRB bearer*

Figure 4.4 describes the operations relating to the DRB bearer.

At transmission, the IP packets are subject to a numbering procedure which allows the receiver to detect duplicate or lost frames and to order.

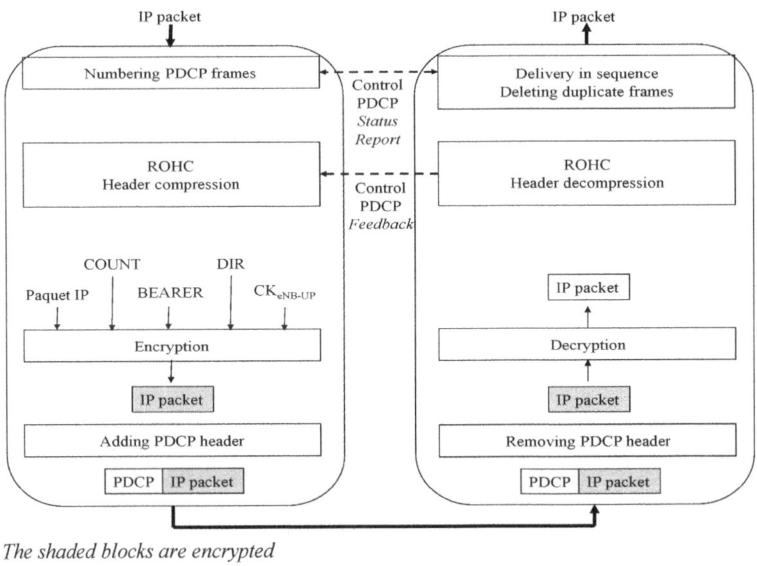

Figure 4.4. *PDCP operations relating to the DRB bearer*

This numbering procedure is used during handover and allows retransmission requests for lost packets, only if the acknowledgment mode is used for the RLC layer.

The IP packet may be subject to the ROHC header compression to reduce the relative weight of headers.

The IP packet is encrypted using the CK_{eNB-UP} key.

The encrypted IP packet is encapsulated by the PDCP header to form the PDCP frame transmitted to the RLC layer.

On receiving the PDCP frames from the RLC layer, the following operations occur:

– the PDCP header is removed;

– the IP packet is decrypted;

– the IP packet is optionally decompressed;

– the duplicate PDCP frames are removed and the IP packets are delivered in sequence.

The PDCP feedback control message checks the header compression procedure.

The PDCP Status Report control message allows recovery of the frames lost during the handover.

4.1.3. *Protocol structure*

The PDCP protocol defines headers to encapsulate the RRC signaling data, the data traffic and the control messages associated with the traffic data.

The structure of the PDCP frames is described in Figure 4.5.

PDCP SN: this field is coded on 5 bits for RRC signaling data and on 7 or 12 bits for traffic data. It indicates the PDCP frame sequence number. This sequence number allows recovery of the PDCP frames lost during the handover.

MAC-I: this field is coded on 4 bytes. It contains the authentication code to control the integrity of the PDCP frame containing RRC signaling data.

Data/Control (D/C): this bit indicates whether the frame contains traffic data (bit value ONE) or control messages specific to the PDCP protocol (bit value ZERO).

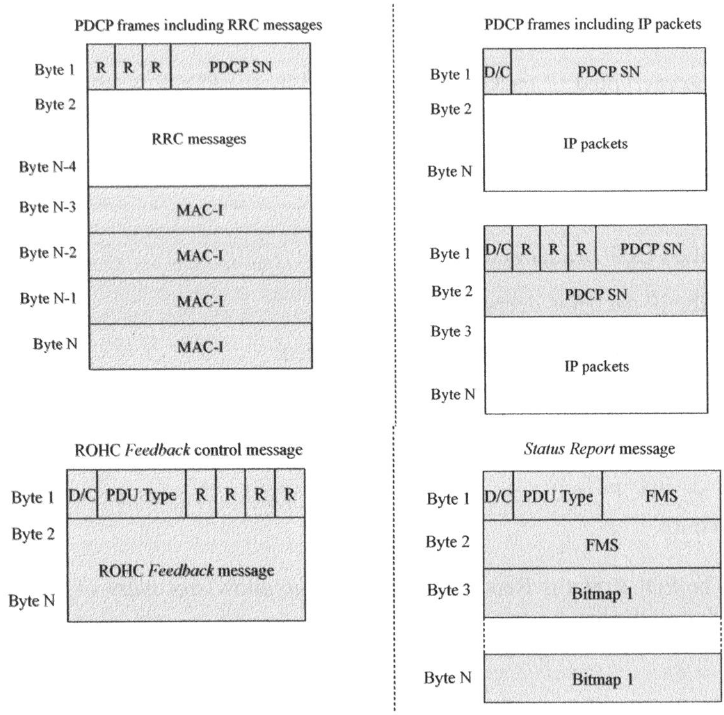

Figure 4.5. *PDCP frame structure*

Packet Data Unit (PDU) type: this field is coded on 3 bits. It shows the control message type associated with traffic data:

– the Status Report (value 000) is used in the re-establishment process after a cell change. It indicates a list of PDCP frame numbers not received during the handover;

– the ROHC feedback (value 001) message is linked to the ROHC header compression mechanism.

First Missing (FMS): this field is coded on 12 bits. It contains the first sequence number of the missing PDCP frames.

Bitmap: this field is a collection of bits indicating whether the PDCP frame was received correctly (bit value ONE) or not (bit value ZERO). The most significant bit of the first byte represents the sequence number following the value of the FMS field.

4.2. RLC protocol

The Radio Link Control (RLC) protocol provides control of the radio link between the mobile and the evolved Node B (eNB) entity.

The mobile can simultaneously activate multiple RLC instances, each instance corresponding to a Packet Data Convergence Protocol (PDCP) instance.

The RLC protocol operates in three operation modes:

– Acknowledged Mode (AM);

– Unacknowledged Mode (UM);

– Transparent Mode (TM) for which no header is added to data.

Figure 4.6 presents the operating modes of the RLC protocol used by the various flow types.

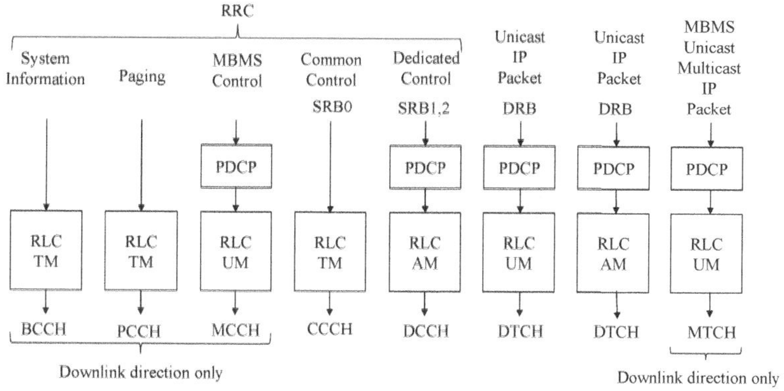

Figure 4.6. *Operating modes of the RLC protocol*

The RLC protocol does the following:

– retransmission in the case of error via the Automatic Repeat reQuest (ARQ) mechanism, for the acknowledged mode only;

– concatenation, segmentation and reassembly of PDCP frames in both the acknowledged and unacknowledged mode;

– possible re-segmentation of PDCP frames, in acknowledgment mode, during a retransmission of the RLC frame;

– re-sequencing of received data, in both the acknowledged and unacknowledged mode;

– detection of duplicate data in both the acknowledged and unacknowledged mode.

4.2.1. *Operations*

4.2.1.1. *TM mode*

At transmission, the Radio Resource Control (RRC) messages are stored in a memory, before being transmitted to the Medium Access Control (MAC) layer.

At reception, the data received from the MAC layer are transmitted directly to the RRC layer.

Figure 4.7 describes the operations performed for the transparent mode.

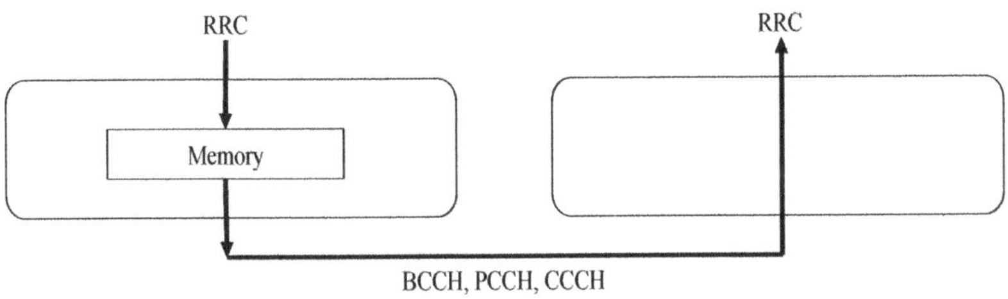

Figure 4.7. *TM mode operations*

4.2.1.2. *UM mode*

At transmission, the PDCP frames are stored in a memory.

When transmitting to the MAC layer, the PDCP frames may be concatenated and/or segmented, and are then encapsulated by an RLC header.

At reception, the data received from the MAC layer are stored in a memory.

The received data are ordered, following desequencing that can be introduced by the the Hybrid Automatic Repeat reQuest (HARQ) mechanism, and the RLC header is then removed.

The last operation is to reassemble the segments to reconstruct the original PDCP frame.

Figure 4.8 describes the operations for the unacknowledged mode.

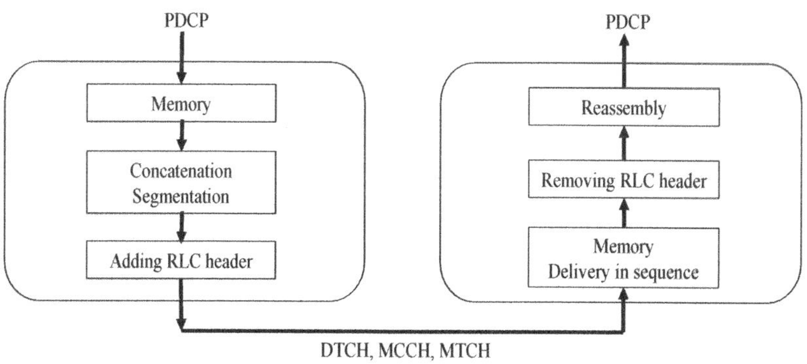

Figure 4.8. *UM mode operations*

4.2.1.3. *AM mode*

At transmission, the PDCP frames are stored in a memory.

When transmitting to the MAC layer, the PDCP frames may be concatenated and/or segmented, and are then encapsulated by an RLC header.

The RLC frames transmitted are stored in a retransmission memory.

The data received from the MAC layer are stored in a memory and are placed in sequence, and the RLC header is then removed.

The last operation is to reassemble the segments to reconstruct the original PDCP frame.

A request may be sent from the source to the destination to receive the status report of an RLC frame reception.

An RLC control message is sent from the destination to the source for the positive or negative acknowledgment of RLC frames.

If the RLC frame is not received correctly, the source will perform a retransmission.

If the Modulation and Coding Scheme (MCS) of the physical layer has been modified, it is possible that the size of the RLC frame is reduced for the retransmission, and in this case, it is necessary to perform a re-segmentation.

Figure 4.9 describes the operations for the acknowledged mode.

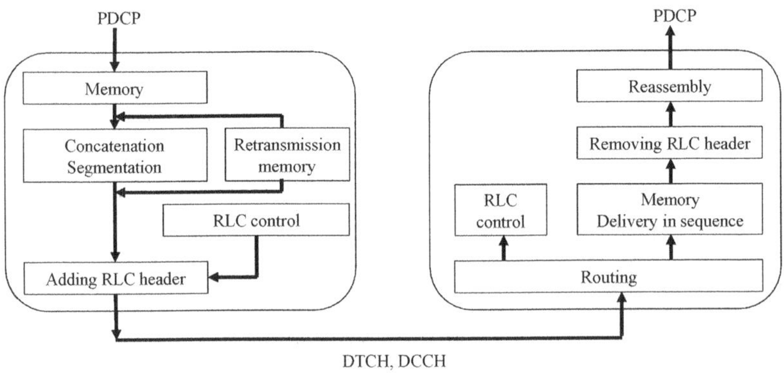

Figure 4.9. *AM mode operations*

4.2.2. *Protocol structure*

The RLC protocol defines a header for the UM mode, a header for the AM mode and a control message.

4.2.2.1. *UM mode*

The header structure depends on the size of the Sequence Number (SN) field and the number of concatenated PDCP frames.

Figure 4.10 describes the RLC header for the following two cases:

– a 5-bit SN field, without concatenation of PDCP frames;

– a 5-bit SN field with concatenation of three PDCP frames without segmentation.

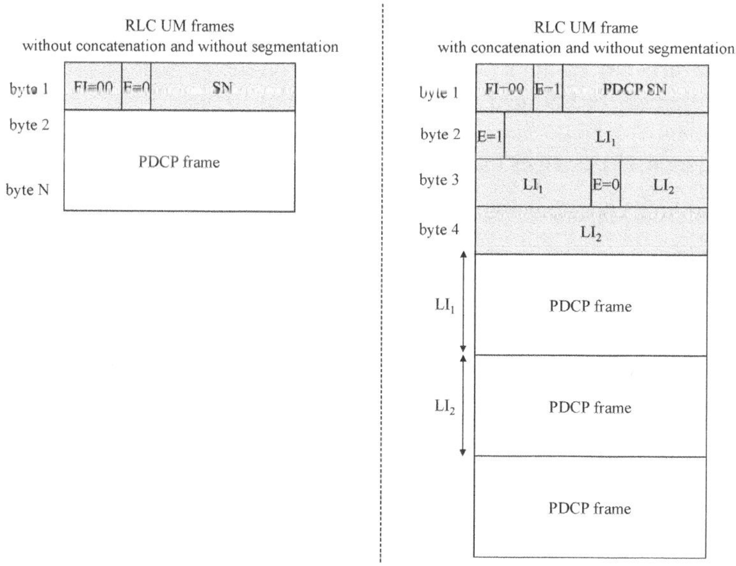

Figure 4.10. *RLC frame structure: UM mode*

Framing Information (FI): this field is coded on 2 bits. It indicates whether or not segmentation is implemented:

– FI = 00, no segmentation;

– FI = 01, start of segmentation of the last PDCP frame;

– FI = 10, end of segmentation of the first PDCP frame;

– FI = 11, end of segmentation of the first PDCP frame and start of segmentation of the last PDCP frame.

Extension (E): this bit indicates the presence (bit value ONE) or absence (bit value ZERO) of an RLC header extension after the next field (SN or LI).

SN: this field is coded on 5 or 10 bits. It contains the RLC frame sequence numbers. The numbering of RLC frames is used to re-sequence the frames received from the MAC sublayer.

Length Indicator (LI): this field is coded on 11 bits. It is only used when concatenating. It contains the size of the corresponding PDCP frame or segment. The LI field is present for each frame, except for the last frame.

4.2.2.2. *AM mode*

The structure of the RLC header depends on the number of concatenated PDCP frames, the SN field having a fixed size of 10 bits.

Figure 4.11 describes the RLC header for the following two cases:

– without concatenation of PDCP frames;

– with segmentation of one PDCP field and concatenation of two PDCP frames.

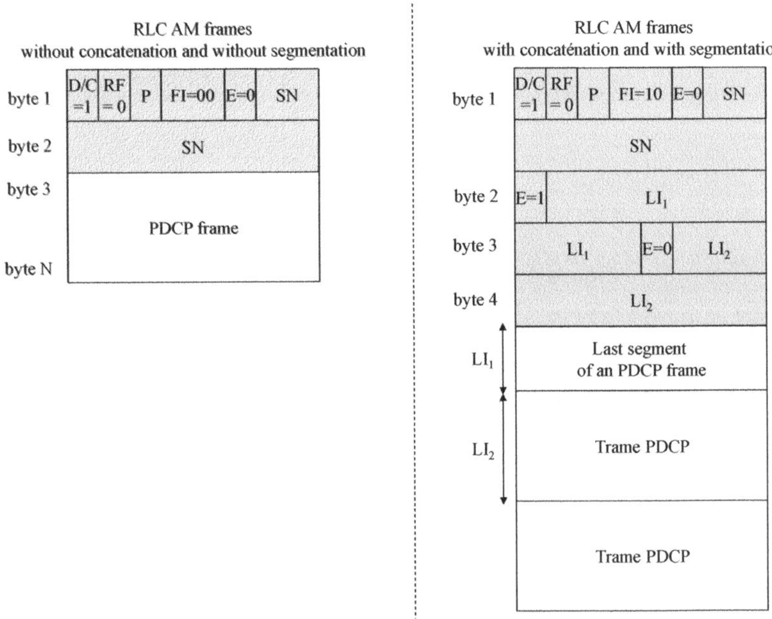

Figure 4.11. *RLC frame structure: initial frame transmission in AM mode*

Data/Control (D/C): this bit indicates whether the frame contains PDCP frames (bit value ONE) or control messages (bit value ZERO).

Resegmentation Flag (RF): this bit indicates whether retransmission is performed with re-segmentation (bit value ONE) or without (bit value ZERO).

Polling (P): this bit indicates whether the transmission of an acknowledgment report for received frames is required (bit value ONE) or not (bit value ZERO).

In the case of retransmission of an RLC frame for which the original resource is not available, the originally transmitted PDCP frame must be able to be segmented.

Figure 4.12 describes the RLC header in the following two cases:

– retransmission of a PDCP frame as a segment;

– retransmission of three PDCP frames, with the first PDCP frame as a segment.

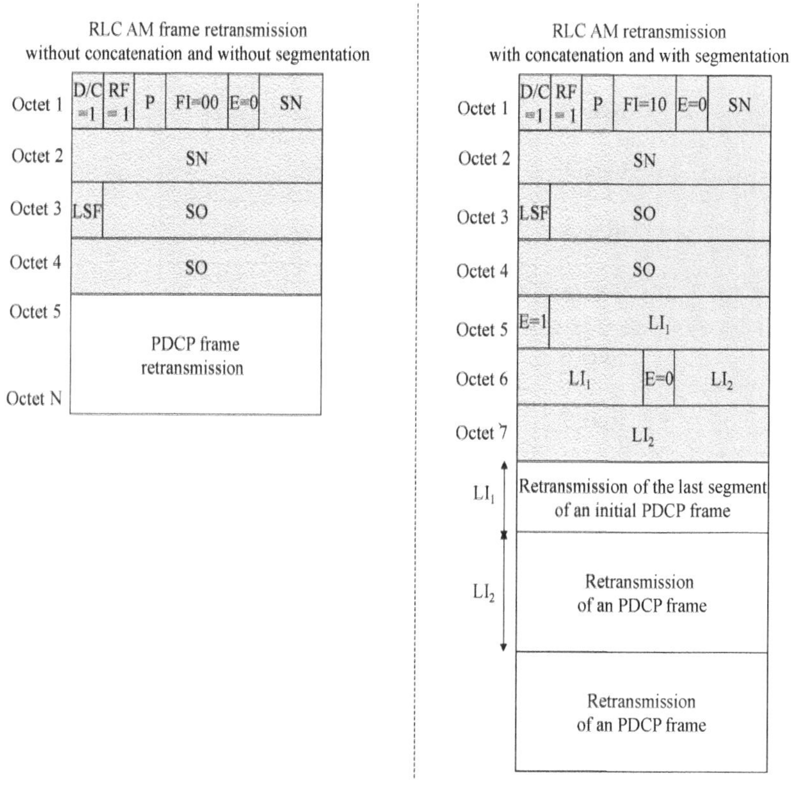

Figure 4.12. *RLC frame structure: frame retransmission in AM mode*

Last Segment Flag (LSF): this bit indicates whether the segment corresponds to the end of the initial PDCP frame (bit value ONE) or not (bit value ZERO).

Segment Offset (SO): this field coded on 15 bits indicates the position of the segment in the original PDCP frame. It contains the current number of the first byte of the segment.

4.2.2.3. *Control message*

The control message concerns the RLC frame acknowledgement and the indication of lost frames and frame segments.

Figure 4.13 describes the RLC protocol control message.

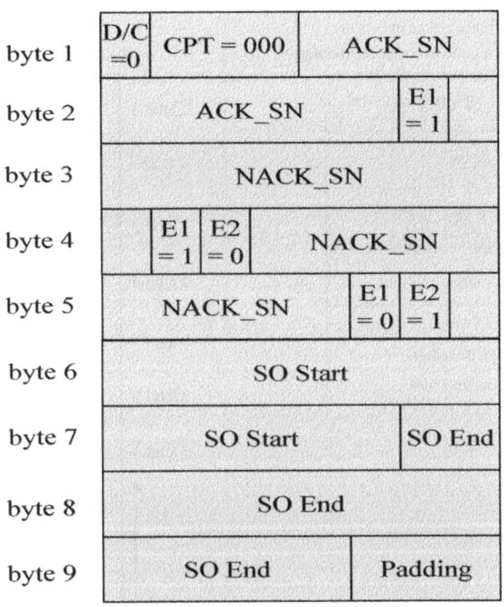

Figure 4.13. *RLC protocol control message*

PDU Control Type (CPT): this field is coded on 3 bits. It indicates the control message type. The only defined message concerns the report relating to the acknowledgement of received RLC frames.

Acknowledgment SN (ACK_SN): this field is coded on 10 bits. It indicates the number of the next expected frame. All transmitted frames, whose number is less than this value, are validated, with the exception of frames whose number is in the NACK_SN field.

Extension 1 (E1): this bit indicates the presence (bit value ONE) or not (bit value ZERO) of NACK_SN, E1 and E2 fields following one another.

Negative Acknowledgment SN (NACK_SN*)*: this field is coded on 10 bits. It contains the number of the lost frame.

Extension 2 (E2): this bit indicates the presence (bit value ONE) or not (bit value ZERO) of SO Start and SO End fields following one another.

SO Start: this field is coded on 15 bits. It indicates the first byte number of the lost segment.

SO End: this field is coded on 15 bits. It indicates the last byte number of the lost segment.

4.3. MAC protocol

The Medium Access Control (MAC) protocol provides the following functions:

– multiplexing Radio Link Control (RLC) frames from multiple instances in a transport block;

– allocation of the resources via a scheduling mechanism for both transmission directions;

– management of retransmission in the case of error via the Hybrid Automatic Repeat reQuest (HARQ) mechanism.

4.3.1. *Operations*

4.3.1.1. *eNB side operations*

The operations at the evolved Node B (eNB) entity level are described in Figure 4.14.

The MAC layer transfers the Paging Control Channel (PCCH) to the Paging Channel (PCH).

The MAC layer transfers the Master Information Block (MIB) messages from the Broadcast Control Channel (BCCH) to the Broadcast Channel (BCH).

The MAC layer performs multiplexing, for the downlink direction, of different logical channels to the Downlink Shared Channel (DL-SCH).

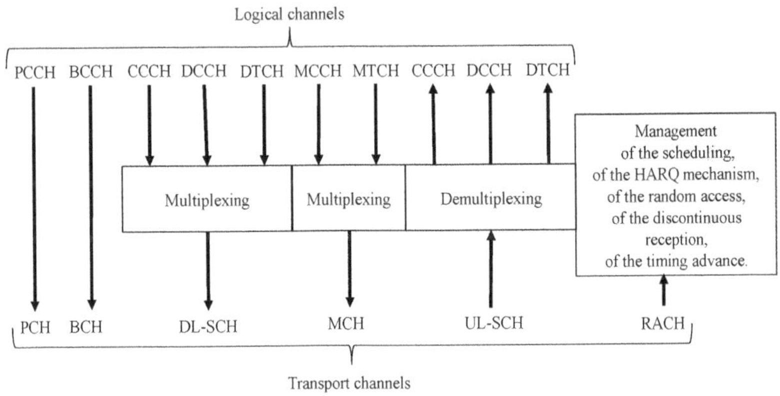

Figure 4.14. *MAC operations: eNB side*

The different logical channels multiplexed for the downlink direction are the following:

– the BCH logical channel, containing System Information Block (SIB) messages;

– the Common Control Channel (CCCH) containing Radio Resource Control (RRC) common control messages;

– the Dedicated Control Channel (DCCH) containing the RRC dedicated control messages;

– the Dedicated Traffic Channel (DTCH) containing Internet Protocol (IP) packets.

The MAC layer performs demultiplexing, for the uplink direction, of the Uplink Shared Channel (UL-SCH) to restore the various logical channels.

The different logical channels resulting from demultiplexing for the uplink direction are the following:

– the CCCH logical channel containing the RRC common control messages;

– the DCCH logical channel containing the RRC dedicated control messages;

– the DTCH logical channel containing the IP packets.

The MAC layer performs multiplexing, for the downlink direction, of the Multicast Control Channel (MCCH) and Multicast Traffic Channel (MTCH) to the Multicast Channel (MCH).

The MAC layer control entity performs the following functions:

– management of the scheduling of the different logical channels for both directions of transmission;

– management of the HARQ retransmission mechanism;

– management of the mobile random access on reception of the Random Access Channel (RACH);

– management of Discontinuous Reception (DRX);

– management of Timing Advance (TA).

4.3.1.2. *UE side operations*

The operations at the User Equipment (UE) level are described in Figure 4.15.

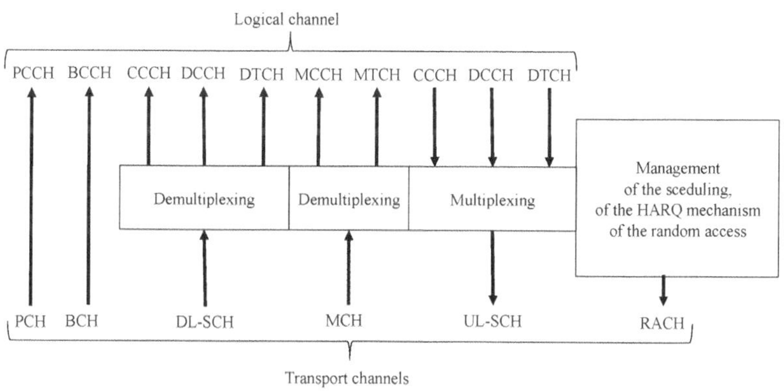

Figure 4.15. *MAC operations: UE side*

The MAC layer transfers the PCH transport channel to the PCCH logical channel.

The MAC layer transfers the MIB messages of the BCH transport channel to the BCCH logical channel.

The MAC layer performs multiplexing, for the uplink direction, of different logical channels to the UL-SCH transport channel.

The different logical channels multiplexed for the uplink direction are as follows:

– the CCCH logical channel containing the RRC common control messages;

– the DCCH logical channel containing the RRC dedicated control messages;

– the DTCH logical channel containing the IP packets.

The MAC layer performs demultiplexing, for the downlink direction, of the DL-SCH transport channel to restore the different logical channels.

The different logical channels resulting from demultiplexing for the downlink direction are as follows:

– the BCH logical channel containing the SIB messages;

– the CCCH logical channel containing the RRC common control messages;

– the DCCH logical channel containing the RRC dedicated control messages;

– the DTCH logical channel containing the IP packets.

The MAC layer performs demultiplexing, for the downlink direction, of the MCH transport channel to restore the MCCH and MTCH logical channels.

The MAC layer control entity performs the following functions:

– management of the scheduling of different logical channels in the resource allocated by the eNB entity;

– management of the HARQ retransmission mechanism;

– generation of the RACH transport channel.

4.3.2. *Protocol structure*

The MAC protocol defines a header for multiplexing RLC frames from different instances and from MAC protocol control elements.

The global header of the MAC protocol therefore consists of a collection of header units, each header unit relating to an RLC frame or a control element.

The order of concatenation of RLC frames or of control elements corresponds to that of the header units.

The headers relating to the control elements precede the headers relating to RLC frames.

Figure 4.16 describes the general MAC frame structure. Each header unit has a size of 2 or 3 bytes depending on the size of the field Length (L).

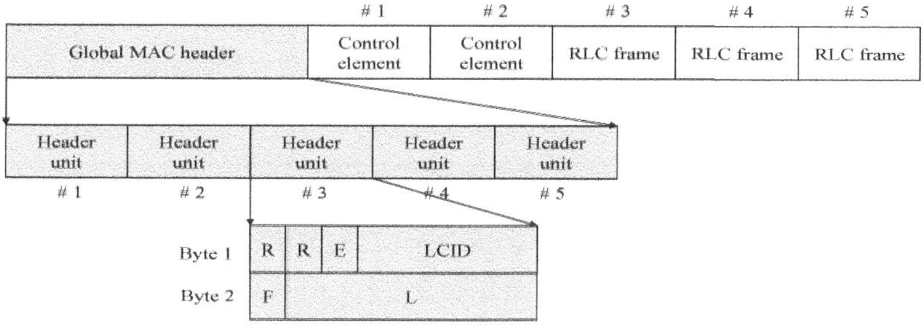

Figure 4.16. *MAC frame structure*

Extension (E): this bit indicates whether the next header unit is present (bit value ONE) or not (bit value ZERO).

Format (F): this bit specifies the Length (L) field format, coded on 7 bits (bit value ONE) or 15 bits (bit value ZERO).

Logical Channel Identifier (LCID): this field is coded on 5 bits. It indicates the logical channel instance identifier or the type of control element (Tables 4.2, 4.3, 4.4).

The identifiers assigned to the DCCH and DTCH logical channels are defined during the connection procedure.

Description	Value
CCCH logical channel identifier	00000.
DCCH and DTCH logical channel identifiers	00001 to 01010
Reserved	01011 to 11010
ADM Activation/Deactivation	11011
UE Contention Resolution Identity (CRI)	11100
Timing Advance (TA)	11101
DRX Discontinuous Reception	11110.
Filling	11111.

The control elements introduced in Release 10 are shaded.

Table 4.2. *LCID field values for the DL-SCH transport channel*

Description	Value
CCCH logical channel identifier	00000.
DCCH and DTCH logical channel identifiers	00001 to 01010
Reserved	01011 to 11000
PHR reserve power available (extended)	11001.
PHR reserve power available	11010.
C-RNTI identifier	11011.
BSR memory status (truncated report)	11100
BSR memory status (short report)	11101
BSR memory status (long report)	11110
Filling	11111

The control elements introduced in Release 10 are shaded.

Table 4.3. *LCID field values for the UL-SCH transport channel*

Description	Value
MCCH logical channel identifier	00000
MTCH logical channel identifier	00001 to 11100
Reserved	11101
MSI scheduling information	11110
Filling	11111

Table 4.4. *LCID field values for the MCH transport channel*

The MAC Random Access Response (RAR) frame is a specific frame used by the eNB entity in response to receiving a random access attempt by the mobile.

As before, the global header contains a collection of header units. The first header may be a specific header providing a delay indication.

Figure 4.17 describes the general structure of a MAC RAR frame with a delay indication.

Extension (E): this bit indicates whether the next header unit is present (bit value ONE) or not (bit value ZERO).

Type (T): this bit provides an indication of the header unit structure (T = 0 for BI, T = 1 for RAPID).

Random Access Preamble Identifier (RAPID): this field coded on 6 bits contains the identifier of the preamble used by the mobile for random access to the eNB entity.

Backoff Indicator (BI): this field coded on 4 bits provides the value of the delay between two consecutive transmissions of the preamble on the Physical Random Access Channel (PRACH).

The RAR message contains the following information:

– Timing Advance (TA) which enables mobile synchronization for the uplink direction;

– allocation of a resource (UL Grant) for the uplink direction so that the mobile can transmit the following RRC messages:

- *ConnectionRequest*,

- *ConnectionRe-establishmentRequest*,

- *ConnectionReconfigurationComplete* during a handover;

– Temporary Cell Radio Network Temporary Identifier (TC-RNTI) allocated to the mobile.

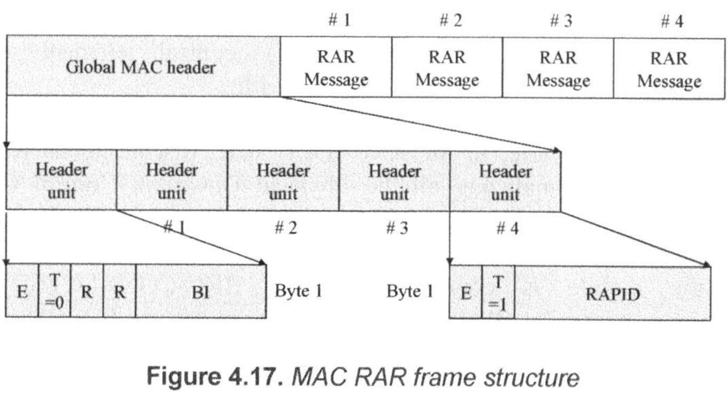

Figure 4.17. *MAC RAR frame structure*

4.3.3. *Control elements*

4.3.3.1. *BSR control element*

The Buffer Status Report (BSR) control element is transmitted by the mobile to provide the memory status in order to obtain the resource for transmission in the Physical Uplink Shared Channel (PUSCH).

The BSR control element can be transmitted periodically, or when new and priority data are stored in the memory or when the size of the MAC frame filling field is greater than that of the control element.

The short and truncated report provides the status of a single memory and contains two fields:

– the Logical Channel Group (LCG) ID field identifies the group of logical channels assigned to that memory;

– the Buffer Size field contains an index corresponding to the size of data available in the memory.

The long report provides the status of all the memories and contains the four Buffer Size fields corresponding to the four logical channel groups.

4.3.3.2. *C-RNTI control element*

The Cell Radio Network Temporary Identifier (C-RNTI) control element contain the identifier assigned to the mobile during random access.

The C-RNTI control element is transmitted by the eNB entity simultaneously with the RRC *ConnectionSetup* message.

4.3.3.3. *DRX control element*

The Discontinuous Reception (DRX) control element enables discontinuous reception to increase the battery life.

If the mobile is idle, in the RRC-IDLE state, discontinuous reception allows it to avoid analyzing all the Physical Downlink Control Channel (PDCCH) to detect the presence of paging.

If the mobile is connected, in the RRC-CONNECTED state, discontinuous reception allows it to avoid analyzing all the PDCCH physical channels to detect the presence of data in the Physical Downlink Shared

Channel (PDSCH) or power commands for the Physical Uplink Control Channel (PUCCH) or for the PUSCH physical channel.

4.3.3.4. *UE CRI control element*

The UE Contention Resolution Identity (CRI) control element is transmitted by the eNB entity simultaneously with the RRC *ConnectionSetup* message.

The UE CRI control element contains either the Shortened Temporary Mobile Subscriber Identity (S-TMSI) if the mobile is attached to the Mobility Management Entity (MME), or a random value otherwise.

As in the allocation of the TC-RNTI identifier, several mobiles can consider themselves to be the holder, the UE CRI control element then resolves the contention.

4.3.3.5. *TA control element*

The Timing Advance (TA) control element contains the timing advance value transmitted by the eNB entity.

The TA control element ensures that data transmitted from several mobiles are synchronized at the eNB entity level.

4.3.3.6. *PHR control element*

The Power HeadRoom (PHR) control element contains the indication of the mobile power reserve, the difference between maximum power and power used for the PUSCH physical channel.

The PHR control element is periodically transmitted by the mobile, the periodicity being indicated in the RRC *ConnectionSetup* or *ConnectionReconfiguration* message transmitted by the eNB entity.

The PHR control element is also transmitted when the variation of the attenuation due to propagation is greater than a threshold indicated in the same RRC messages.

Release 10 introduced a new control element that is used to indicate the power reserve for each of the aggregated Component Carrier (CC).

4.3.3.7. *MSI control element*

The MCH Scheduling Information (MSI) control element indicates scheduling information of the Multicast Channel (MCH).

The MSI control element indicates the end of MTCH physical channel if the MCH transport channel allocated more data than the MTCH physical channel requires.

4.3.3.8. *ADM control element*

The Activation/Deactivation MAC (ADM) control element was introduced in Release 10 and concerns activation and deactivation of the SCell secondary radio channels.

The ADM control element is used when radio channels have been previously established by an RRC *ConnectionSetup* or *ConnectionReconfiguration* message.

The control element saves on the consumption of the mobile, rapid deactivation of an SCell radio channel allowing the mobile to avoid processing relating to that channel.

5

Physical Layer

5.1. Frequency plan

5.1.1. *Frequency bands*

The medium for the two directions of transmission uses two matching bandwidths in the Frequency Division Duplex (FDD) mode or a single bandwidth in the Time Division Duplex (TDD) mode.

For the FDD mode, each direction of transmission functions simultaneously in the allocated bandwidth.

For the TDD mode, the two directions of transmission function in the same bandwidth, each direction is allocated a certain amount of time.

Table 5.1 shows the frequency bands which are used for the radio interface in FDD mode.

The frequency bands allocated in uplink direction have a value lower than those allocated in downlink direction, except for bands 13, 14, 20 and 24, in order to obtain better propagation conditions.

Band 6 was removed from the frequency plan in Release 9 and 10 and replaced with band 19.

Number	Uplink frequency band	Downlink frequency band
1	1920-MHz–1980-MHz	2110-MHz–2170-MHz
2	1850-MHz–1910-MHz	1930-MHz–1990-MHz
3	1710-MHz–1785-MHz	1805-MHz–1880-MHz
4	1710-MHz–1755-MHz	2110-MHz–2155-MHz
5	824-MHz–849-MHz	869-MHz–894MHz
6	830-MHz–840-MHz	875-MHz–885-MHz
7	2500-MHz–2570-MHz	2620-MHz–2690-MHz
8	880-MHz–915-MHz	925-MHz–960-MHz
9	1749.9-MHz–1784.9-MHz	1844.9-MHz–1879.9-MHz
10	1710-MHz–1770-MHz	2110-MHz–2170-MHz
11	1427.9-MHz–1447.9-MHz	1475.9-MHz–1495.9-MHz
12	699-MHz–716-MHz	729-MHz–746-MHz
13	777-MHz–787-MHz	746-MHz–756-MHz
14	788-MHz–798-MHz	758-MHz–768-MHz
17	704-MHz–716-MHz	734-MHz–746-MHz
18	815-MHz–830-MHz	860-MHz–875-MHz
19	830-MHz–845-MHz	875-MHz–890-MHz
20	832-MHz–862-MHz	791-MHz–821-MHz
21	1447.9-MHz–1462.9-MHz	1495.9-MHz–1510.9-MHz
22	3410-MHz–3490-MHz	3510-MHz–3590-MHz
23	2000-MHz–2020-MHz	2180-MHz–2200-MHz
24	1626.5-MHz–1660.5-MHz	1525-MHz–1559-MHz
25	1850-MHz–1915-MHz	1930-MHz–1995-MHz

Table 5.1. *Frequency plan: FDD mode*

Table 5.2 shows the frequency bands which are used for the radio interface in TDD mode.

Number	Frequency band
33	1900-MHz–1920-MHz
34	2010-MHz–2025-MHz
35	1850-MHz–1910-MHz
36	1930-MHz–1990-MHz
37	1910-MHz–1930-MHz
38	2570-MHz–2620-MHz
39	1880-MHz–1920-MHz
40	2300-MHz–2400-MHz
41	2496-MHz–2690-MHz
42	3400-MHz–3600-MHz
43	3600-MHz–3800-MHz

Table 5.2. *Frequency plan: TDD mode*

Bands 33 to 40 were defined in Release 8.

Bands 41 to 43 were introduced in Release 10.

5.1.2. *Radio channel*

The bandwidth of the radio channel is flexible and can take several values.

The radio channel is made of an Orthogonal Frequency Division Multiplexing (OFDM) of subcarriers.

The step between subcarriers is equal to 15 kHz or 7.5 kHz for the downlink direction and equal to 15 kHz for the uplink direction.

Table 5.3 shows the total and the useful values of the bandwidth of the radio channel and the number subcarriers with 15 kHz spacing.

Bandwidth of the radio channel	1.4-MHz	3-MHz	5-MHz	10-MHz	15-MHz	20-MHz
Useful bandwidth	1.08-MHz	2.7-MHz	4.5-MHz	9-MHz	13.5-MHz	18-MHz
Number of subcarriers	72	180	300	600	900	1200

Table 5.3. *Bandwidth of the radio channel*

The same frequency band can be used in each covered cell by the eNB entity. In order to avoid the interference, the following plan is set up (Figure 5.1):

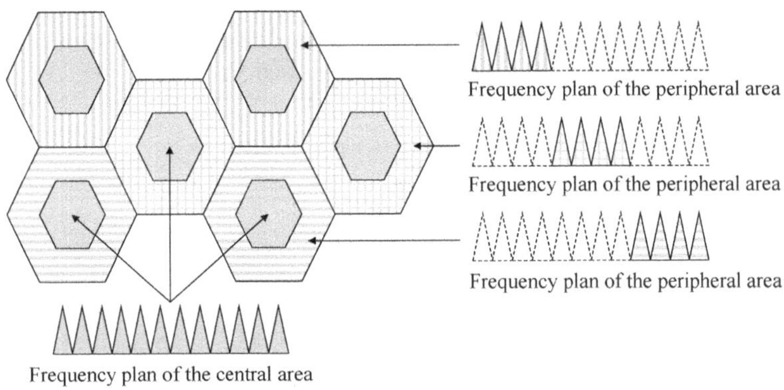

Figure 5.1. *Division of the radio channel between the cells*

– the whole spectrum is used in the central area of the cell, which has lower interference created by the neighboring cells;

– a part of the spectrum is used in the peripheral area of the cell, which facilitates the removal of interference from the neighboring cells. The Inter-Cell Interference Coordination (ICIC) mechanism facilitates the dynamic division of the spectrum between the cells, depending on traffic.

The indicator of the charge of the cell is provided by the X2-AP LOAD INFORMATION message, which may contain the following information:

– the Interference Overload Indication (IOI) information is related to the interference detected by the eNB entity, for the uplink direction. The eNB entity receiving this information must reduce the level emitted by the mobile;

– the High Interference Indication (HII) information is related to the interference detected by the eNB entity, for the uplink direction, by showing the impacted bandwidth. The eNB entity receiving this information must avoid using the indicated bandwidth, for the uplink direction, and those for the mobiles located in the periphery of the cell;

– the Relative Narrowband Tx Power (RNTP) information is related to the reduction of the power produced by an eNB entity. The eNB entity receiving this information includes it into its mechanism of the traffic scheduling.

5.1.3. *Aggregation of the radio channels*

The Carrier Aggregation (CA) of the radio channels was introduced in Release 10.

The aggregation of the radio channels involves combining the use of several Component Carrier (CC) in order to increase the throughput of the cell.

The aggregation, whose maximum value of the bandwidth is 100 MHz, can occur on five radio channels.

The radio channels can be aggregated according to several patterns (Figure 5.2):

– the radio channels can be contiguous in the same frequency bands;

– the radio channels can be non-contiguous in the same frequency bands;

– the radio channels can be localized in different frequency bands.

The notation related to the aggregation of radio channels determines the frequency band and the class of the bandwidth of the radio channel (Figure 5.2).

Class A corresponds to a bandwidth of the radio channel less or equal to 20 MHz.

Class C corresponds to a bandwidth of the radio channel less or equal to 40 MHz.

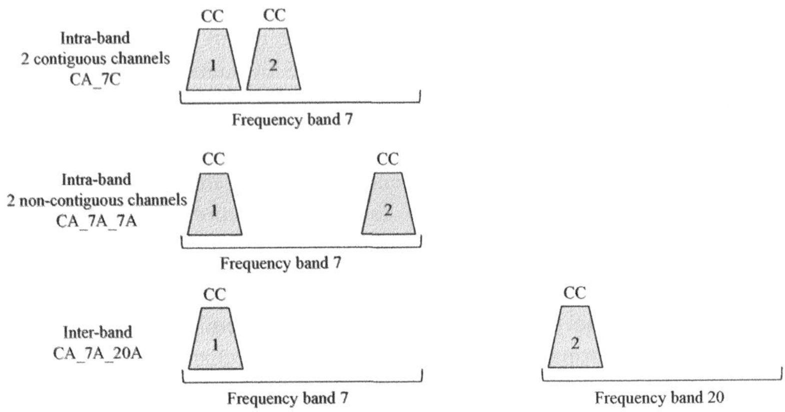

Figure 5.2. *Radio channel aggregation*

One of the radio channels makes up the primary cell PCell, and it has the following characteristics:

– the mechanisms for random access and for RRC connections only occur on the primary cell PCell;

– the Non-Access Stratum (NAS) messages related to the mobility and the session management are only transmitted in the primary cell PCell.

The other radio channels are made of the secondary cells SCell.

5.1.4. *Numbering of radio channels*

The center frequency of the radio channel is determined starting from the E-UTRA Absolute Radio Frequency Channel Number (EARFCN), which can be of a value between 0 and 65535.

The relationship between the EARFCN number (N_{DL}) and the center frequency of the radio channel (F_{DL}), for the downlink direction, is provided by the following formula:

$$F_{DL} = F_{DL\_low} + 0.1(N_{DL} - N_{Offs\text{-}DL})$$

The relationship between EARFCN number (N_{UL}) and the center frequency of the radio channel (F_{UL}), for the uplink direction, is provided by the following formula:

$$F_{UL} = F_{UL\_low} + 0.1(N_{UL} - N_{Offs\text{-}UL})$$

Table 5.4 provides the parameters for the calculation of the center frequency of the radio channel, for the downlink direction and for the FDD mode.

Number	F_{DL_low}	$N_{Offs\text{-}DL}$	N_{DL}
1	2110	0	0–599
2	1930	600	600–1199
3	1805	1200	1200–1949
4	2110	1950	1950–2399
5	869	2400	2400–2649
6	875	2650	2650–2749
7	2620	2750	2750–3449
8	925	3450	3450–3799
9	1844.9	3800	3800–4149
10	2110	4150	4150–4749
11	1475.9	4750	4750–4949
12	729	5010	5010–5179
13	746	5180	5180–5279
14	758	5280	5280–5379
17	734	5730	5730–5849
18	860	5850	5850–5999
19	875	6000	6000–6149
20	791	6150	6150–6449
21	1495.9	6450	6450–6599
22	3510	6600	6600–7399
23	2180	7500	7500–7699
24	1525	7700	7700–8039
25	1930	8040	8040–8689

Table 5.4. *Calculation parameters of the center frequency, for the downlink direction, for the FDD mode*

Table 5.5 provides the parameters for the calculation of the center frequency of the radio channel, for the uplink direction and for the FDD mode.

Number	F_{UL_low}	$N_{Offs\text{-}UL}$	N_{UL}
1	1920	18000	18000–18599
2	1850	18600	18600–19199
3	1710	19200	19200–19949
4	1710	19950	19950–20399
5	824	20400	20400–20649
6	830	20650	20650–20749
7	2500	20750	20750–21449
8	880	21450	21450–21799
9	1749.9	21800	21800–22149
10	1710	22150	22150–22749
11	1427.9	22750	22750–22949
12	699	23010	23010–23179
13	777	23180	23180–23279
14	788	23280	23280–23379
17	704	23730	23730–23849
18	815	23850	23850–23999
19	830	24000	24000–24149
20	832	24150	24150–24449
21	1447.9	24450	24450–24599
22	3410	24600	24600–25399
23	2000	25500	25500–25699
24	1626.5	25700	25700–26039
25	1850	26040	26040–26689

Table 5.5. *Calculation parameters of the center frequency, for the uplink direction for the FDD mode*

Table 5.6 provides the parameters for the calculation of the center frequency of the radio channel, for the TDD mode.

Number	F_{DL_low} F_{UL_low}	$N_{Offs\text{-}DL}$ $N_{Offs\text{-}UL}$	N_{DL} N_{UL}
33	1900	36000	36000–36199
34	2010	36200	36200–36349
35	1850	36350	36350–36949
36	1930	36950	36950–37549
37	1910	37550	37550–37749
38	2570	37750	37750–38249
39	1880	38250	38250–38649
40	2300	38650	38650–39649
41	2496	39650	39650–41589
42	3400	41590	41590–43589
43	3600	43590	43590–45589

Table 5.6. *Calculation parameters of the center frequency for the TDD mode*

5.2. Multiplexing structure

5.2.1. *Time-division multiplexing*

Two structures of time frames are defined according to the Frequency Division Duplex (FDD) or Time Division Duplex (TDD) mode.

For the uplink direction, the signals transmitted by the different mobiles must be temporally aligned, to the reception by the evolved Node B (eNB) entity.

The mobiles must therefore be synchronized temporally by the eNB entity which conveys the Timing Advance (TA) to them, to be applied for the uplink direction.

5.2.1.1. *Structure of the type-1 frame*

The type-1 structure defined for the FDD mode lasts 10 ms and contains 10 subframes (Figure 5.3).

Each subframe is made up of two time slots.

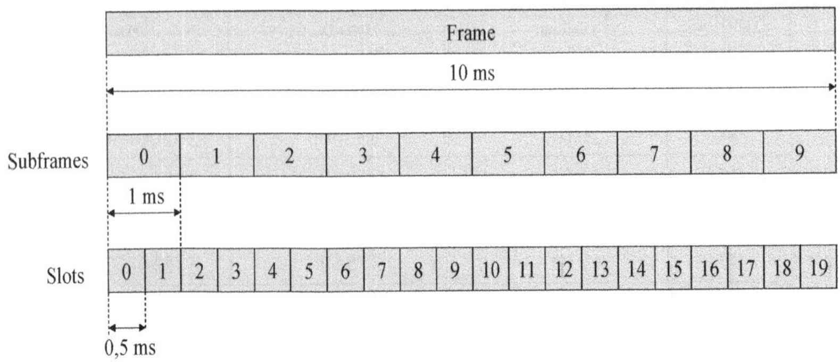

Figure 5.3. *Structure of the frame in FDD mode*

5.2.1.2. *Structure of the type-2 frame*

The type-2 structure defined for the TDD mode also lasts 10 ms and contains two semi-frames of 5 ms each (Figure 5.4).

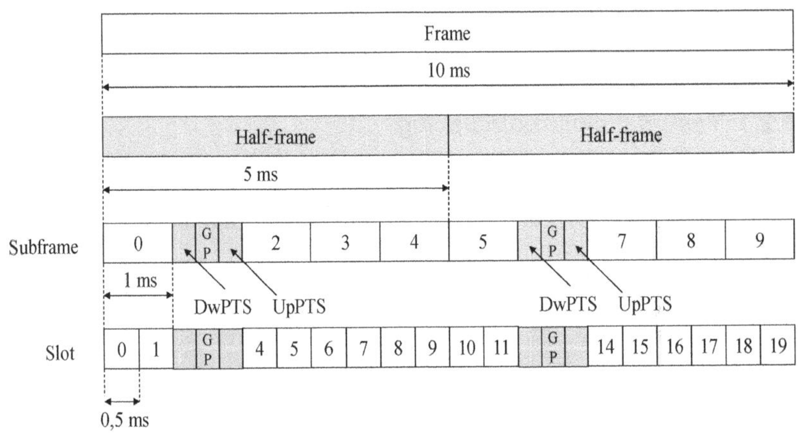

Figure 5.4. *Structure of the frame in TDD mode*

Each half-frame is comprised of 5 subframes and the second can correspond to a special subframe containing three particular fields:

– a field for the Downlink Pilot Time Slot (DwPTS) in the downlink direction. This field can contain data;

– a field for the Uplink Pilot Time Slot (UpPTS) in the uplink direction. This field can contain data or a preamble;

– a field of Gap Period (GP), between the two preceding fields. This interval time facilitates the compensation of a time difference between different mobiles and avoids an overlap between the two transmission directions.

The subframes are attributed to the data for the uplink and downlink direction according to diverse configurations (Table 5.7 and Figure 5.5):

– subframes 0 and 5 are always allocated to traffic in the downlink direction;

– subframe 1 is always allocated to the special subframe containing the three particular fields;

– subframe 2 is always allocated to traffic in the uplink direction;

– subframe 6 can be allocated to the special subframe containing three particular fields for a periodicity of 5 ms;

– subframes 3, 4, 7, 8, 9 are allocated to the data for the downlink or uplink direction according to the selected configuration.

Configuration	Periodicity	Number of the subframe									
		0	1	2	3	4	5	6	7	8	9
0	5 ms	D	S	U	U	U	D	S	U	U	U
1	5 ms	D	S	U	U	D	D	S	U	U	D
2	5 ms	D	S	U	D	D	D	S	U	D	D
3	10 ms	D	S	U	U	U	D	D	D	D	D
4	10 ms	D	S	U	U	D	D	D	D	D	D
5	10 ms	D	S	U	D	D	D	D	D	D	D
6	5 ms	D	S	U	U	U	D	S	U	U	D

D subframe attributed to the downlink direction
U subframe attributed to the uplink direction
S special subframe containing the three particular fields

Table 5.7. *Configuration of the TDD frame*

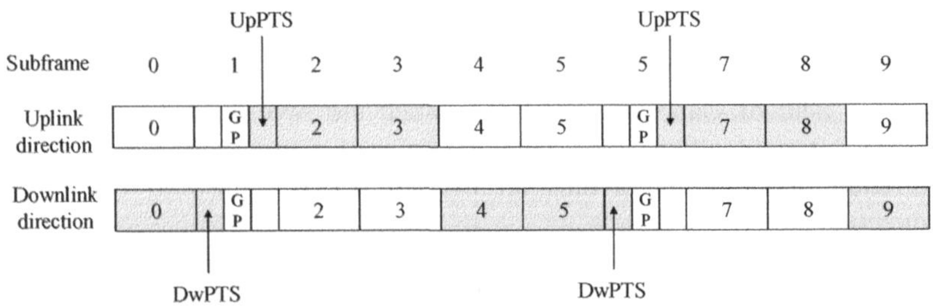

Grey shows subframes allocated to each transmission direction

Figure 5.5. *Structure of the TDD frame: configuration 1*

The structure of the special subframe is described in Table 5.8.

Special subframe configuration	Normal cyclic prefix for the downlink direction			Extended cyclic prefix for the downlink direction		
	DwPTS	UpPTS		DwPTS	UpPTS	
		Normal cyclic prefix	Extended cyclic prefix		Normal cyclic prefix	Extended cyclic prefix
0	6592*Ts			7680*Ts		
1	19760*Ts			20480*Ts	2192*Ts	2560*Ts
2	21952*Ts	2192*Ts	2560*Ts	23040*Ts		
3	24144*Ts			25600*Ts		
4	26336**Ts*			7680*Ts		
5	6592*Ts			20480*Ts	4384*Ts	5120*Ts
6	19760*Ts			23040*Ts		
7	21952*Ts	4384*Ts	5120*Ts	–	–	–
8	24144*Ts			–	–	–

$T_s = 1/(15000 \times 2048)$ seconds

Table 5.8. *Structure of the special subframe*

5.2.1.3. *Structure of the time slot*

Each time slot is comprised of 3 or 6 or 7 Orthogonal Frequency-Division Multiplexing (OFDM) symbols (Figure 5.6).

A symbol corresponds to a number of bits which depends on the modulation used:

– 2 bits in the case of a Quadrature Phase-Shift Keying (QPSK) modulation whose constellation contains 4 different symbols;

– 4 bits in the case of a 16-Quadrature Amplitude Modulation (16-QAM) whose constellation contains 16 different symbols;

– 6 bits in the case of a 64-QAM modulation whose constellation contains 64 different symbols.

The introduction of a guard time between the OFDM symbols facilitates the removal of inter-symbol interference.

The guard time of a time slot is found at the beginning of the OFDM symbol. It contains the copy of the end of the symbol in order to avoid a very high dynamic for the amplifiers. This copy is called Cyclic Prefix (CP) (Figure 5.6).

The normal cyclic prefix is used in case the delay between the different reflected signals is low, which is the case in small diameter cells (Figure 5.6 and Table 5.9).

The extended cyclic prefix is essentially used in case the delay between the different reflected signals is significant, which is the case in cells of a large diameter (Figure 5.6 and Table 5.9).

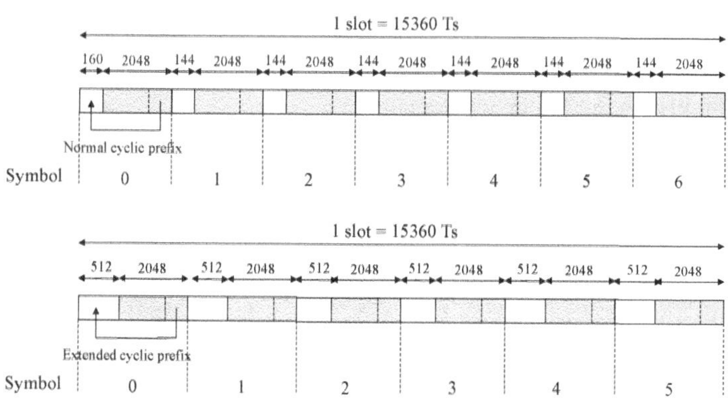

$T_s = 1/(15000 \times 2048)$ seconds

Figure 5.6. *Structure of the time slot*

Downlink direction		Length of the cyclic prefix
Normal cyclic prefix	Δf = 15 kHz	160 * Ts for $l = 0$ 144 * Ts for $l = 1,2,...,6$
Extended cyclic prefix	Δf = 15 kHz	512 * Ts for $l = 0,1,...,5$
	Δf = 7.5 kHz	1024 * Ts µs for $l = 0,1,2$
Uplink direction		Length of the cyclic prefix
Normal cyclic prefix		160 * Ts for $l = 0$ 144 * Ts for $l = 1,2,...,6$
Extended cyclic prefix		512 * Ts for $l = 0,1,...,5$

Δf: step between subcarriers
l: number of the symbol in the time slot

Table 5.9. *Length of the cyclic prefix*

5.2.2. *Resource block*

A Resource Element (RE) is the smallest unit which can be attributed to a signal (Figure 5.7)

The resource element corresponds to an OFDM symbol in the time domain and to a subcarrier in the frequency domain.

The Resource Block (RB) is the resource unit attributed to a mobile (Figure 5.7).

The resource block corresponds to 0.5 ms (1 time slot) in the time domain and to 180 kHz in the frequency domain.

The resource element is identified by the pair (k,l), k represents the subcarrier index in the frequency domain and l the OFDM symbol index in the time domain:

$k = 0,..., N_{\text{RB}}^{\text{DL}} N_{\text{sc}}^{\text{RB}} - 1$ and $l = 0,..., N_{\text{symb}}^{\text{DL}} - 1$ for the downlink direction

$k = 0,..., N_{\text{RB}}^{\text{UL}} N_{\text{sc}}^{\text{RB}} - 1$ and $l = 0,..., N_{\text{symb}}^{\text{UL}} - 1$ for the uplink direction

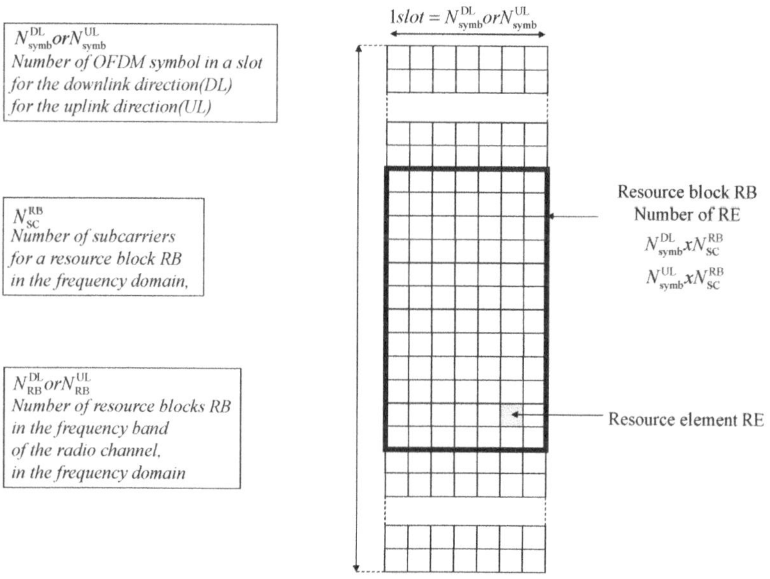

Figure 5.7. *Structure of the resource block*

The number of resource blocks (N^{DL}_{RB} *or* N^{UL}_{RB}) in the frequency domain depends on the bandwidth of the radio channel (Table 5.10).

Bandwidth (MHz)	1.4	3	5	10	15	20
Useful bandwidth(MHz)	1.08	2.7	4.5	9	13.5	18
Number of resource blocks	6	15	25	50	75	100

Table 5.10. *Structure of the radio channel for the frequency domain*

The number of subcarriers and of OFDM symbols per resource block depends on the spacing between the subcarriers and the size of the cyclic prefix (Table 5.11).

For a normal cyclic prefix, a resource block contains 84 resource elements (7 OFDM symbols × 12 subcarriers) (Table 5.11).

For an extended cyclic prefix, a resource block contains 72 resource elements (6 OFDM symbols × 12 subcarriers or 3 OFDM symbols × 24 subcarriers) (Table 5.11).

<table>
<tr><td colspan="2">Downlink direction</td><td>N_{sc}^{RB}</td><td>N_{symb}^{DL}</td></tr>
<tr><td>Normal cyclic prefix</td><td>Δf = 15 kHz</td><td rowspan="2">12</td><td>7</td></tr>
<tr><td rowspan="2">Extended cyclic prefix</td><td>Δf = 15 kHz</td><td>6</td></tr>
<tr><td>Δf = 7,5 kHz</td><td>24</td><td>3</td></tr>
<tr><td colspan="2">Uplink direction</td><td>N_{sc}^{RB}</td><td>N_{symb}^{UL}</td></tr>
<tr><td colspan="2">Normal cyclic prefix</td><td>12</td><td>7</td></tr>
<tr><td colspan="2">Extended cyclic prefix</td><td>12</td><td>6</td></tr>
</table>

Table 5.11. *Parameters of the resource block*

5.2.3. *Resource element group*

The Resource Element Group (REG) is a specific structure used by the control channel of the physical layer:

– Physical HARQ Indicator Channel (PHICH);

– Physical Control Format Indicator Channel (PCFICH);

– Physical Downlink Control Channel (PDCCH).

The control channel of the physical layer can comprise of up to the first three OFDM symbols of each subframe.

The resource element group is comprised of 4 resource elements represented by the index (k',l'):

$k' = k$, k corresponds to the index of the subcarrier of the resource element group with the lowest value.

l' corresponds to the number of the OFDM symbol in the time slot.

The number of resource element groups in a resource block, depends on the number of resource element reserved to the Cell-Specific Reference Signal (RS).

For the first OFDM symbol of the subframe, it is possible to create two resource element groups in a resource block. We systematically reserved four resource elements for the Cell-Specific RS physical signal. This corresponds to the use of two antenna ports (Figure 5.8).

For the second OFDM symbol of the subframe, two cases can be found (Figure 5.8):

– three resource element groups are created in a resource block. This corresponds to the use of two antenna ports;

– two resource element groups are created in a resource block. This corresponds to the use of four antenna ports.

For the third OFDM symbol, three resource element groups are created in a resource block, no matter how many antenna ports there are (Figure 5.8).

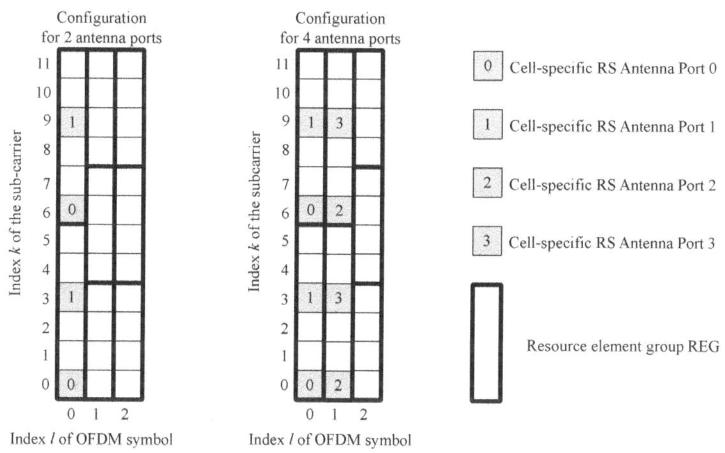

Figure 5.8. *Structure of the resource element group*

5.3. Transmission chain

The data on the interface between the physical layer and the Medium Access Control (MAC) is in the form of transport blocks or of control information of the physical layer.

The transmission chain is comprised of two subsets, whose interface is made of the physical channel:

– for each transmission direction, the first subset is comprised of the detection and error correction codes and of the rate matching;

– for the downlink direction, the second subset is comprised of the modulation, of the mapping on spatial layers, of the precoding, the mapping on the resource elements and of the Inverse Fast Fourier Transform (IFFT) for generating the Orthogonal Frequency Division Multiple Access (OFDMA) signal (Figure 5.9);

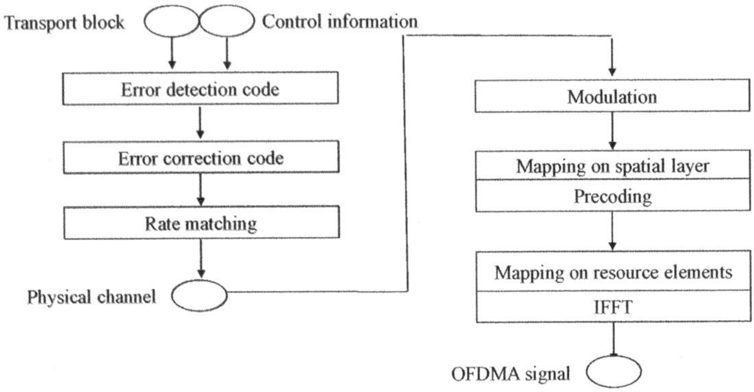

Figure 5.9. *Transmission chain: downlink direction*

– for the uplink direction, the second subset is comprised of the modulation, of the mapping on the resource elements and of the inverse fast Fourier transform IFFT. The generation of the Single Carrier Frequency Division Multiple Access (SC-FDMA) signal introduces a Fast Fourier Transform (FFT). The mapping on the spatial layers and the precoding are only implemented for Release 10 (Figure 5.10).

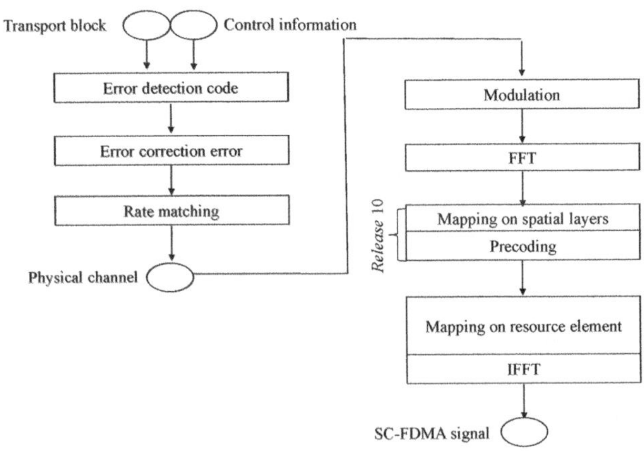

Figure 5.10. *Transmission chain: uplink direction*

The mapping of transport blocks and of control information on the physical channel is described in Table 5.12.

Transport blocks	Physical channels
UL-SCH	PUSCH
RACH	PRACH
DL-SCH	PDSCH
PCH	PDSCH
BCH	PBCH
MCH	PMCH
Control information	Physical channels
DCI	PDCCH
CFI	PCFICH
HI	PHICH
UCI	PUCCH, PUSCH

Table 5.12. *Mapping on the physical channels*

5.3.1. *Error detection codes*

The error detection code is provided by a Cyclic Redundancy Check (CRC) code.

The CRC structure is the remainder of the data block division to be transmitted by a generator polynomial; the remainder is made of cyclic redundancy bits.

If the size of the set made up of the transport block and of the bits of cyclic redundancy is larger than 6144 bits, this set is divided into code blocks and each segment must have its own error detection code.

The error detection code at the level of the receptor, allows for the detection of the residual errors, which in turn enables the triggering of the Hybrid Automatic Repeat reQuest (HARQ) mechanism.

Table 5.13 provides an overview of the different error detection codes used by the physical channels.

Physical channels	Error detection code
PUSCH	Transport block: CRC 24A Segment: CRC 24B
PUCCH	CRC 8
PDSCH	Transport block: CRC 24A Segment: CRC 24B
PBCH	CRC 16
PMCH	Transport block: CRC 24A Segment: CRC 24B
PDCCH	CRC16

Table 5.13. *Error detection codes*

5.3.2. Error correction codes

Table 5.14 provides an overview of different error correction codes used by the physical channels.

Physical channels	Error correction code
PUSCH	Turbo code
PUCCH	Code per block Convolution code
PDSCH	Turbo code
PBCH	Convolution code
PMCH	Turbo code
PDCCH	Convolution code
PCFICH	Code per block
PHICH	Repetition code

Table 5.14. *Error correction codes*

The rate matching retains a certain number of redundancy bits generated by the coding and determines the coding rate which is the ratio between the size of blocks before and after the error correction code.

The coding scheme defines the value of the coding rate to be applied.

The coding scheme used for the Physical Downlink Shared Channel (PDSCH) and Physical Uplink Shared Channel (PUSCH) is determined by the MAC layer and they depend on the conditions of radio propagation and on the number of bits to be transmitted.

The coding scheme used for the Physical Multicast Channel (PMCH) is a configured value in the Multi-cell/Multicast Coordination Entity (MCE).

5.3.3. *Modulation*

In the case of a Binary Phase Shift Keying (BPSK) modulation, a bit is mapped in the complex symbol $x=I+jQ$. The constellation is comprised of 2 symbols (Figure 5.11).

In the case of a Quadrature Phase Shift Keying (QPSK) modulation, two bits are mapped in the complex symbol $x=I+jQ$. The constellation is comprised of 4 symbols (Figure 5.11).

In the case of a 16-Quadrature Amplitude Modulation (16-QAM), four bits are mapped in the complex symbol $x=I+jQ$. The constellation is comprised of 16 symbols (Figure 5.11).

In the case of a 64-QAM modulation, six bits are mapped in the complex symbol $x=I+jQ$. The constellation is comprised of 64 symbols (Figure 5.11).

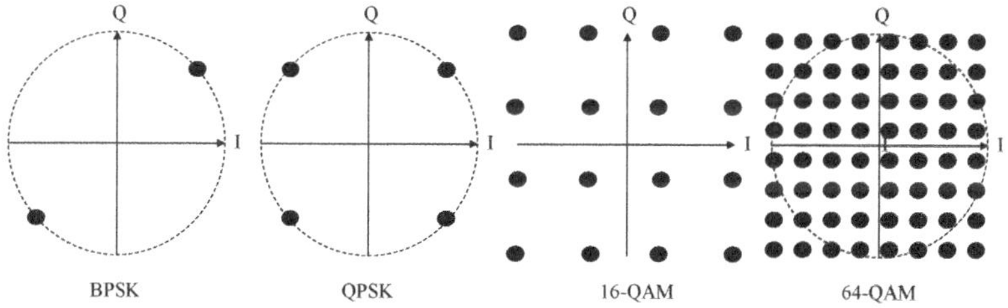

Figure 5.11. *Different symbol constellations*

Table 5.15 provides an overview of the different modulations used by the physical channels.

The modulation scheme used for the PDSCH and PUSCH physical channels is determined by the MAC layer and it depends on radio propagation conditions and on the number of bits to be transmitted.

Physical channels	Modulation
PUSCH	QPSK, 16-QAM, 64-QAM (1)
PUCCH	BPSK, QPSK
PDSCH	QPSK, 16-QAM, 64-QAM
PBCH	QPSK
PMCH	QPSK, 16-QAM, 64-QAM
PDCCH	QPSK
PCFICH	QPSK
PHICH	BPSK

(1) The availability of the 64 QAM modulation depends on the category of the mobile.

Table 5.15. *Modulations*

The modulation scheme used for the PMCH physical channel is a value configured in the MCE entity.

5.3.4. *Antenna ports*

The symbols produced by the modulation are scattered on the spatial layers (mapping function), and then precoded.

Precoded symbols are associated with some Reference Signals (RS) involving antenna ports.

Table 5.16 shows the association of antenna ports and of reference signals for the downlink direction.

Antenna port	Release	Reference signal
p0 to p3	8	Cell-Specific RS
p4	8	MBSFN RS
p5	8	UE-Specific RS
p6	9	Positioning RS
p7 and p8	9	UE-Specific RS
p9 to p14	10	UE-Specific RS
p15 to p22	10	CSI RS

Table 5.16. *Association of the antenna ports and reference signals, for the downlink direction*

For antenna ports p0 to p3, each antenna port is associated with a physical antenna.

Antenna port p4 is associated with a single physical antenna.

Antenna port p5 is associated with two or four physical antennas.

Antenna port p6 is associated with a single physical antenna.

For antenna ports p7 to p14 and p15 to p22, each antenna port is associated with a single physical antenna.

Table 5.17 shows the association of antenna ports and physical channels for the uplink direction.

The antenna port number depends on the antenna port number identified by the $\tilde{p}$ index.

Physical channel or signal	$\tilde{p}$ index	Antenna ports		
		1 port	2 ports	4 ports
PUSCH SRS	0	10	20	40
	1		21	41
	2			42
	3			43
PUCCH	0	100	200	
	1		201	

Table 5.17. *Numbering of antenna ports for the uplink direction*

Antenna ports 10 and 100 use a single physical antenna $\tilde{p} = 0$.

Antenna ports 20 and 200 (21 and 201 respectively) use the physical antenna $\tilde{p} = 0$ ($\tilde{p} = 1$ respectively).

For antenna ports 40 to 43, each antenna port uses a physical antenna.

5.3.5. *Transmission modes*

5.3.5.1. *Downlink direction*

Transmission modes 1 to 7 were defined in Release 8.

Transmission mode 8 was defined in Release 9.

Transmission mode 9 was defined in Release 10.

Transmission mode 1 is the Single Input Single Output (SISO) type. It corresponds to setting a transmitter to antenna port 0.

When the mobile is equipped with two receptors, it can implement the diversity reception.

Transmission mode 2 is the Multiple Input Single Output (MISO) type. It corresponds to setting several transmitters of the same signal to antenna ports p0 and p1 (when there are two transmitters) or p0 to p3 (when there are four transmitters) to support the transmit diversity function which facilitates the improvement of the quality of the signal received.

In the case of transmission on the two antenna ports p0 and p1, the transmit diversity corresponds to the Space Frequency Block Coding (SFBC) mechanism.

In the case of transmission to the four antenna ports p0 to p3, the transmit diversity corresponds to SFBC/SFTD (Frequency Shift Transmit Diversity) mechanism.

Transmission mode 3 is the Multiple Input Multiple Output (MIMO) type. It corresponds to setting two transmitters of different signals to antenna ports p0 and p1 (2×2 MIMO) or four transmitters to antenna ports p0 to p3 (4×4 MIMO) to support the spatial multiplexing function in an open loop which facilitates the improvement of the throughput of the cell.

Transmission mode 4 is the 2×2 MIMO or 4×4 MIMO type. It corresponds to setting two transmitters of different signals to antenna ports p0 and p1 (2×2 MIMO) or setting four transmitters to antenna ports p0 to p3 (4×4 MIMO) to support the spatial multiplexing function in a closed loop.

Transmission mode 5 is the multi-user MIMO type. It supports the spatial multiplexing function in a closed loop, of two (2×2 MIMO) or four users (4×4 MIMO).

Transmission mode 6 corresponds to a simplified version of transmission mode 4 for which a single spatial layer is used.

Transmission mode 7 supports the beamforming function. This mode uses the p5 antenna port.

Transmission mode 7 can also simultaneously support the multi-user MIMO function, for providing the spatial multiplexing of several users in an open loop, with each user granted one layer.

The formation of beams in the direction of the mobile facilitates the increase of the cell range. To obtain this result, each signal emitted on an antenna port is dephased in a way that it can receive all phased signals at the mobile.

Transmission mode 8 is an extension of transmission mode 7. The spatial multiplexing allows for the following configurations:

– two users with an allocation of two spatial layers per user;

– four users with an allocation of a spatial layer per user.

Transmission mode 8 uses the antenna ports p7 and p8.

Transmission mode 9 is the 8×8 MIMO type. It also supports the beamforming function and the multi-user spatial multiplexing.

Transmission mode 9 is associated with the antenna ports p7 to p14 in the case of a MIMO mono-user.

Table 5.18 recapitulates the different transmission modes. The eNB entity is able to commute between several transmission techniques without signaling by a Radio Resource Control (RRC) message.

5.3.5.2. *Uplink direction*

For Release 8 and 9, there is no defined transmission mode due to the use of a single antenna port.

Since the mobile is equipped with two antennas, the transmission can be carried out on one of the two antennas.

The antenna selection is either carried out through the mobile, or through the eNB entity.

Mode	Transmission scheme
1	Antenna port 0
2	Transmit diversity
3	Transmit diversity
	MIMO in open loop
4	Transmit diversity
	MIMO in closed loop
5	Transmit diversity
	MIMO multi-users
6	Transmit diversity
	MIMO in closed loop, a single layer
7	Transmit diversity or antenna port 0
	Beam formation, antenna port 5
8	Transmit diversity or antenna port 0
	Antenna ports 7 and 8
9	Transmit diversity or antenna port 0
	MIMO, antenna ports 7 to 14

Table 5.18. *Transmission modes for the downlink direction*

If the eNB entity is equipped with two receptors, it can implement the diversity reception.

Release 10 introduces transmission modes 1 and 2.

Transmission mode 1 is the SISO type. It corresponds to setting a transmitter to antenna port p10 for the PUSCH physical channel or to antenna port p100 for the Physical Uplink Control Channel (PUCCH).

Transmission mode 2 is the MIMO type for the PUSCH physical channel. It corresponds to setting two transmitters of different signals on the antenna ports p20 and p21 (2x2 MIMO) or four transmitters of different signals on the antenna port p40 to p43 (4x4 MIMO) to support the spatial multiplexing function in a closed loop.

Transmission mode 2 uses transmit diversity for the PUCCH physical channel. It corresponds to setting two transmitters of the same signal on antenna ports p200 and p201.

5.3.6. *Inverse Fast Fourier Transform*

The OFDMA signals for the downlink direction and SC-FDMA for the uplink direction are generated starting from an Inverse Fast Fourier Transform.

For the downlink direction , the IFFT processes the precoded and mapped symbols on the resource elements in order to go from the frequency domain to the time domain.

The OFDMA term also defines the resource which the mobile accessed. One allocates the time (1 ms corresponding to a subframe) and the frequency (blocks of 180 kHz corresponding to 12 subcarriers with 15 kHz spacing or to 24 subcarriers with 7.5 kHz spacing) to the mobile.

Frequency division multiplexing of subcarriers use the Orthogonal Frequency Division Multiplexing (OFDM) mechanism.

For the uplink direction, before carrying out mapping on the resource elements, one applies a Fast Fourier Transform (FFT) to the symbols.

This operation aims to reduce the Peak Average Power Ratio (PAPR) between the peak power and the average power of the signal emitted by the mobile, in order to reduce its energy consumption.

The combination of the FFT and IFFT transforms allows for reconstruction of a mono-carrier frequency, therefore the term SC-FDMA is used for the uplink direction.

One also allocates time and frequency to the mobile, as for the downlink direction.

5.3.7. *Configuration of the transmission*

The transmission architecture must support the three configuration types of aggregation of radio channels.

Configuration A corresponds to a contiguous channel aggregation (Figure 5.12).

The transmission architecture includes a single chain comprising of the inverse fast Fourier transform, of the Digital Analog Converter (DAC), of the mixer, of the Power Amplifier (PA) and of the filtering.

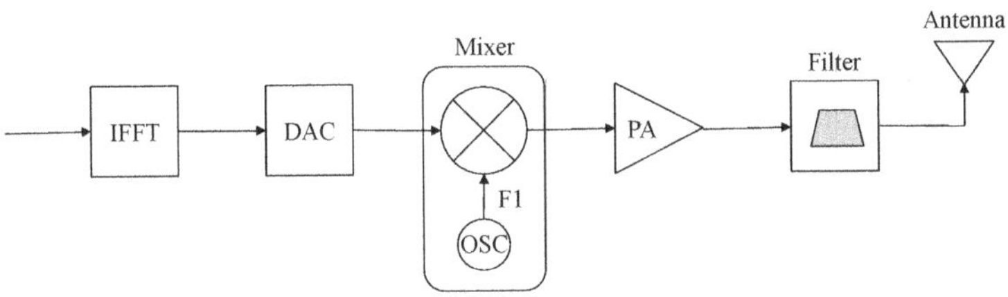

Figure 5.12. *Configuration of the transmission: aggregation of two contiguous channels*

Configuration B corresponds to a non-contiguous channel aggregation in the same frequency band (Figure 5.13).

The transmission architecture comprises of two chains, each comprising of the IFFT, of the digital analog converter and of the mixer.

The two chains are then joined to be amplified, then filtered.

Configuration C corresponds to an aggregation of two localized channels in two different frequency bands (Figure 5.14).

The transmission architecture includes two independent chains, each comprising of the Invesse Fast Foulier Transform, of the digital analog converter, of the mixer, of the power amplifier and of the filtering.

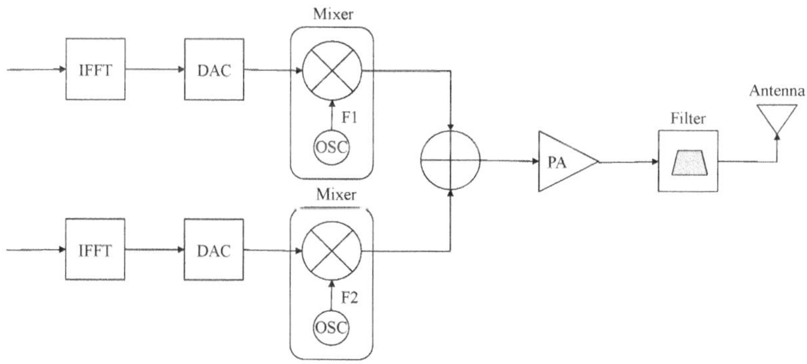

Figure 5.13. *Configuration of the transmission: aggregation of two non-contiguous channels in the same frequency band*

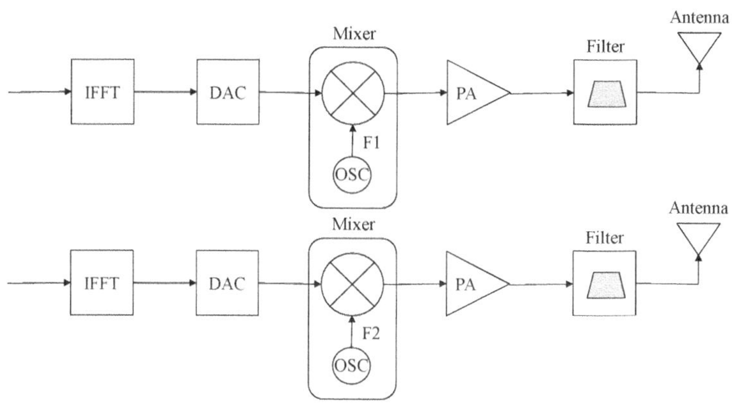

Figure 5.14. *Configuration of the transmission: aggregation of two channels in two different frequency bands*

5.3.8. *Antenna configurations*

The performance of each transmission mode depends on the characteristics of antenna correlation.

The transmission modes implementing the beamforming and the multi-user MIMO needs a good correlation between the antennas.

The transmission modes implementing the transmit diversity and the mono-user MIMO requires a decorrelation between the antennas.

For an antenna made of columns of radiating component with vertical polarization, the correlation between the columns is relatively strong.

For an antenna made of a column of two sets of radiating components, each set corresponding to a crossed polarization ± 45 degrees, the correlation is relatively weak.

The antenna configurations are described in Figure 5.15 and relate to a single frequency band.

Configuration A corresponds to an antenna made of a radiating component column with vertical polarization.

Configuration B corresponds to an antenna made of two radiating component columns with vertical polarization.

Configuration C corresponds to an antenna made of four radiating component columns with vertical polarization.

Configuration D corresponds to an antenna made of a column of two sets of radiating components, each set corresponding to a crossed polarization ± 45 degrees.

Configuration E corresponds to an antenna made of two columns, each column comprising of two radiating components sets, each set corresponding to a crossed polarization ± 45 degrees.

Configuration E combines pairs of correlated radiating components of the same polarization, and of pairs of decorrelated radiating components with different polarization.

Configuration F corresponds to an antenna made:

– of a central column of two sets of radiating components, each set ensemble corresponds to a crossed polarization ± 45 degrees;

– of two separate sets, each set is comprised of radiating components, corresponding to one of the two crossed polarizations ± 45 degrees. The remaining polarization remains available for use in another frequency band.

Configuration G corresponds to an antenna made of four columns of radiating components, each column is comprised of two sets of radiating components, each set corresponds to a crossed polarization ± 45 degrees.

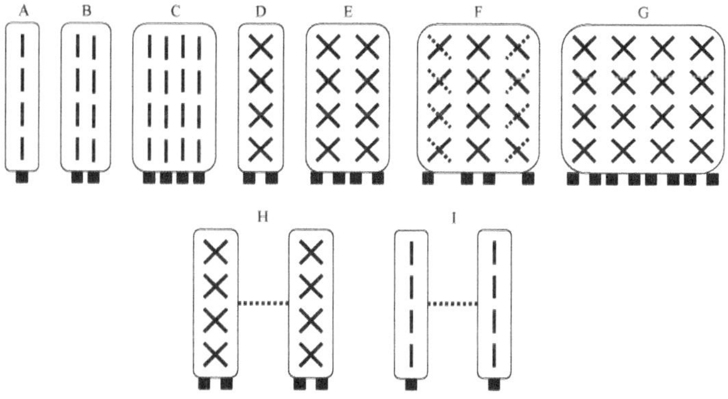

Figure 5.15. *Antenna configurations*

Configuration H corresponds to two separate antennas, each antenna is made of a column of two sets of radiating components, each set corresponds to a crossed polarization ± 45 degrees.

Configuration I corresponds to two separate antennas, each antenna is made of a column of radiating components with vertical polarization.

Table 5.19 defines the transmission mode adapted for each antenna configuration.

Antenna configuration	A	B	C	D	E	F	G	H	I
Transmission mode	1	5, 7	5 7-8	2-4 6	2-6 7-8	2-6	2-6 7-8 9	2-6	2-4 6

Table 5.19. *Correlation between the antenna configuration and the transmission mode*

6

Downlink Physical Signals

6.1. PSS physical signal

The Primary Synchronization Signal (PSS) performs the following functions:

– frequency synchronization;

– time synchronization, at the level of the Orthogonal Frequency Division Multiplexing (OFDM) symbol, of the time slot, of the subframe (1 ms periodicity) and of the half-frame (5 ms periodicity);

– value determination for the number $N_{\mathrm{ID}}^{(2)}$.

The number $N_{\mathrm{ID}}^{\mathrm{cell}}$ for the Physical-layer Cell Identity (PCI) equals $N_{\mathrm{ID}}^{\mathrm{cell}} = 3N_{\mathrm{ID}}^{(1)} + N_{\mathrm{ID}}^{(2)}$.

$N_{\mathrm{ID}}^{(1)}$ represents the group number and can take values ranging from 0 to 167. It is determined by the Secondary Synchronization Signal (SSS).

$N_{\mathrm{ID}}^{(2)}$ represents the number in the group and can take values ranging from 0 to 2.

6.1.1. *Sequence generation*

The PSS physical signal is generated from a 62-symbol Zadoff-Chu sequence $d(n)$ corresponding to the following equations:

$$d_u(n) = \begin{cases} e^{-j\frac{\pi un(n+1)}{63}} & n = 0,1,...,30 \\ e^{-j\frac{\pi u(n+1)(n+2)}{63}} & n = 31,32,...,61 \end{cases}$$

where u represents the root index, it being possible for u to adopt the three values shown in Table 6.1.

$N_{\text{ID}}^{(2)}$	u
0	25
1	29
2	34

Table 6.1. *Root index numbers for the PSS physical signal*

6.1.2. *Mapping on the resource elements*

The mobile should be able to analyze the PSS physical signal without knowing the bandwidth of the radio channel.

The sequence $d(n)$ is mapped on 62 subcarriers either side of the subcarrier corresponding to the central frequency of the radio channel.

Ten additional subcarriers located either side of the central subcarrier are reserved.

The ten additional subcarriers and the central subcarrier do not support any information.

The mapping of the sequence $d(n)$, in the time domain, is determined by the type of time frame.

For the case of the type-1 time frame, corresponding to the Frequency Division Duplex (FDD), the $d(n)$ sequence is mapped on the last OFDM symbol of time slots 0 and 10 (Figure 6.1).

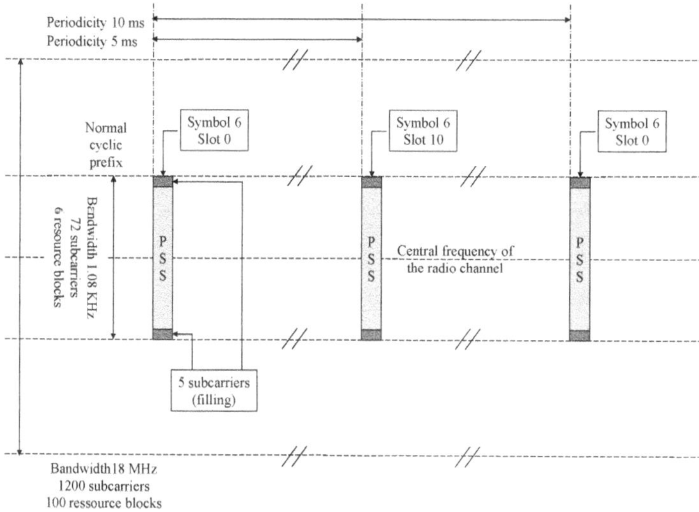

Figure 6.1. *Mapping of the PSS physical signals: FDD mode*

For the case of the type-2 time frame, corresponding to the Time Division Duplex (TDD) mode, the $d(n)$ sequence is mapped on the third OFDM symbol of time slots 2 and 12 (Figure 6.2).

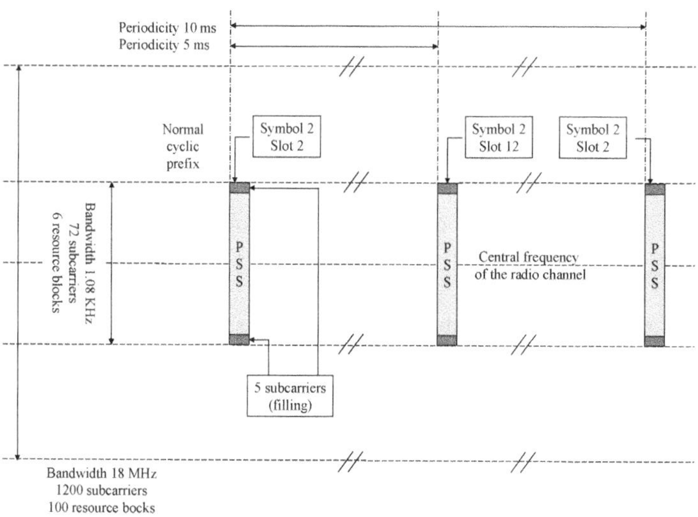

Figure 6.2. *Mapping of the PSS physical signals: TDD mode*

The PSS physical signal is transmitted through the downlink direction regardless of the type of configuration of the TDD mode.

The PSS physical signal is transmitted through subframe 1, which contains Downlink Pilot Time Slot (DwPTS) symbols, and through subframe 6, which contains either the DwPTS symbols (configurations 0, 1, 2, 6) or a downlink direction (configuration 3, 4, 5).

6.2. SSS physical signal

The Secondary Synchronization Signal (SSS) performs the following functions:

– time synchronization at the frame level (10 ms periodicity);

– determination of the use of either Frequency Division Duplex (FDD) mode or of Time Division Duplex (TDD) mode;

– determination of the type of Cyclic Prefix (CP) to be used, normal or extended;

– determination of the value of number $N_{\text{ID}}^{(1)}$.

The determination of the FDD or TDD mode, on the one hand, and the type of cyclic prefix on the other hand, can be accomplished by means of temporal spacing between the Primary Synchronization Signal (PSS) and the SSS physical signal.

The number $N_{\text{ID}}^{\text{cell}}$ for the Physical-layer Cell Identity (PCI) equals $N_{\text{ID}}^{\text{cell}} = 3N_{\text{ID}}^{(1)} + N_{\text{ID}}^{(2)}$.

$N_{\text{ID}}^{(1)}$ represents the group number and can take values ranging from 0 to 167.

$N_{\text{ID}}^{(2)}$ represents the number inside the group and can take values ranging from 0 to 2. It is determined by the PSS physical signal.

6.2.1. *Sequence generation*

The SSS physical signal corresponds to a sequence $d(n)$, composed of two sequences, an even rank sequence $d(2n)$ and an odd rank sequence $d(2n+1)$, interleaved and obtained by means of the following formulas:

– subframe 0: $$\begin{aligned} d(2n) &= s_0^{(m_0)}(n)c_0(n) \\ d(2n+1) &= s_1^{(m_1)}(n)c_1(n)z_1^{(m_0)}(n) \end{aligned} \quad \text{with } 0 \le n \le 30$$

– subframe 5: $$\begin{aligned} d(2n) &= s_1^{(m_1)}(n)c_0(n) \\ d(2n+1) &= s_0^{(m_0)}(n)c_1(n)z_1^{(m_1)}(n) \end{aligned} \quad \text{with } 0 \le n \le 30$$

Both sequences $s_0^{(m_0)}(n)$ and $s_1^{(m_1)}(n)$ are pseudo-random sequences with 31 symbols.

m_0 and m_1 represent the cyclic shift of the sequence. Their value depends on the value of number $N_{\mathrm{ID}}^{(2)}$.

Both sequences $s_0^{(m_0)}(n)$ and $s_1^{(m_1)}(n)$ are scrambled by two sequences $c_0(n)$ and $c_1(n)$ that depend on the value of the number $N_{\mathrm{ID}}^{(2)}$.

The scrambled pseudo-random sequence of the odd rank is scrambled again by two sequences $z_1^{(m_0)}(n)$ and $z_1^{(m_1)}(n)$.

6.2.2. *Mapping on the resource elements*

The sequence $d(n)$ is mapped on 62 subcarriers located on either side of the subcarrier corresponding to the central frequency of the radio channel.

Ten additional subcarriers located on either side of the central subcarrier are reserved.

The ten additional subcarriers and the central subcarrier do not support any information.

The mapping of the sequence $d(n)$ depends on the type of time frame.

For the case of the type-1 time frame, corresponding to the FDD mode, the sequence $d(n)$ is mapped to the second last Orthogonal Frequency Division Multiplexing (OFDM) of time slots 0 and 10 (Figure 6.3).

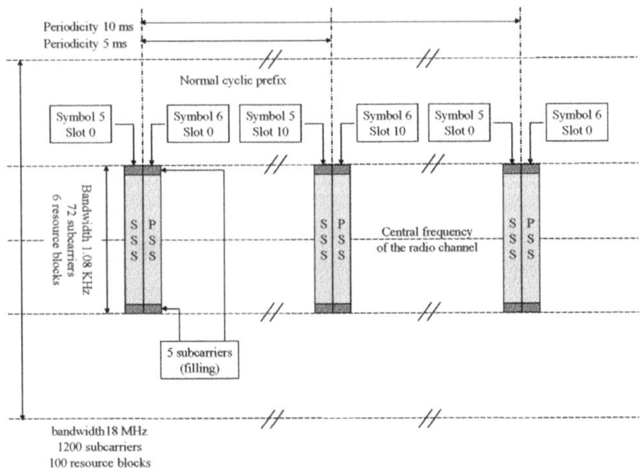

Figure 6.3. *Mapping of the SSS physical signal: FDD mode*

For the case of the type-1 time frame, the PSS and SSS physical signals are adjacent.

For the case of the type-2 time frame, corresponding to the Time Division Duplex (TDD) mode, the sequence $d(n)$ is mapped to the last OFDM symbol of time slots 1 and 11 (Figure 6.4).

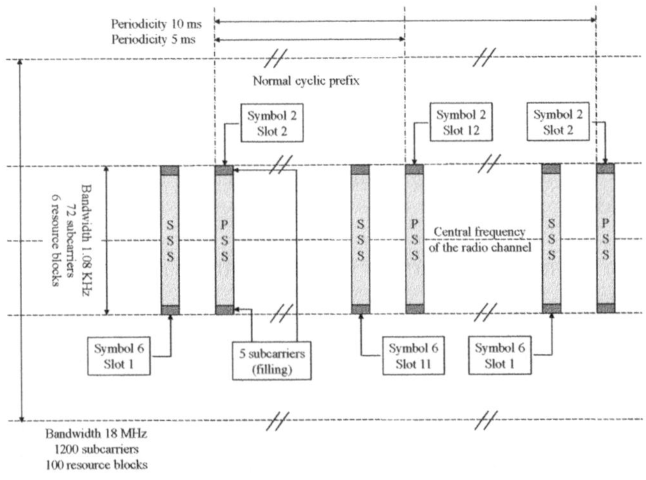

Figure 6.4. *Mapping of the SSS physical signals: TDD mode*

For the case of the type-2 time frame, the PSS and SSS physical signals are not adjacent.

6.3. Cell-Specific RS physical signal

The Cell-Specific Reference Signal (RS) is used to perform the coherent demodulation of the signal.

The coherent demodulation of the received signal is based on calculating the transfer function of the radio channel.

The Cell-Specific RS physical signal allows for the implementation of spatial multiplexing and for the transmit diversity.

The Cell-Specific RS physical signal makes it possible to measure the Reference Signal Received Power (RSRP) and the Reference Signal Received Quality (RSRQ).

The Cell-Specific RS physical signal is associated with the following antenna ports:

– antenna port p0, in the case of transmission on one antenna;

– antenna ports p0 and p1, in the case of transmission on two antennas;

– antenna ports p0, p1, p2 and p3, in the case of transmission on four antennas.

The Cell-Specific RS physical signal is only used when the step along subcarriers is equal to 15 kHz.

The Cell-Specific RS physical signal is transmitted on each subframe and covers the entire bandwidth of the radio channel.

6.3.1. *Sequence generation*

The Cell-Specific RS physical signal is obtained through the following sequence:

$$r_{l,n_s}(m) = \frac{1}{\sqrt{2}}\left(1-2\cdot c(2m)\right) + j\frac{1}{\sqrt{2}}\left(1-2\cdot c(2m+1)\right)$$

$$m = 0,1,\ldots,2N_{\text{RB}}^{\text{max,DL}} - 1$$

$N_{\mathrm{RB}}^{\mathrm{max,\,DL}}$ is expressed as the number of resource blocks in the frequency domain, for the entire bandwidth of the radio channel.

l corresponds to the number of the Orthogonal Frequency Division Multiplexing (OFDM) symbol of the time slot.

n_{s} corresponds to the time slot number of the time frame.

$c(m)$ corresponds to a pseudo-random sequence that is 31-bit long.

At the start of every OFDM symbol, the initial c_{init} value of the pseudo-random sequence is obtained with the following formula:

$$c_{\mathrm{init}} = 2^{10} \cdot \left(7 \cdot \left(n_{\mathrm{s}} + 1\right) + l + 1\right) \cdot \left(2 \cdot N_{\mathrm{ID}}^{\mathrm{cell}} + 1\right) + 2 \cdot N_{\mathrm{ID}}^{\mathrm{cell}} + N_{\mathrm{CP}}$$

$N_{\mathrm{ID}}^{\mathrm{cell}}$ corresponds to the Physical-layer Cell Identity (PCI).

N_{CP} takes the value of ONE for a normal cyclic prefix and of ZERO for an extended cyclic prefix.

6.3.2. *Mapping on the resource elements*

For a normal cyclic prefix and for antenna ports p0 and p1, the mapping takes place on the first and the fifth OFDM symbol of each time slot.

The resource elements used for the Cell-Specific RS physical signal are spaced at 6 subcarrier intervals in the frequency domain.

The Cell-Specific RS physical signal can take six positions in the frequency domain; each position depends on the value $N_{\mathrm{ID}}^{\mathrm{cell}}$ of the cell physical identity.

The resource elements that are used for the fifth OFDM symbol are shifted from three subcarriers, in the frequency domain, as compared with those used for the first OFDM symbols.

The resource elements that are used for the antenna port p1 are shifted from three subcarriers, in the frequency domain, as compared with those used for the port p0.

The resource elements used for the antenna port p0 must not be used for the antenna port p1, and vice versa.

For a normal cyclic prefix and for antenna ports p2 and p3, the mapping takes place on the second OFDM symbol of each time slot (Figure 6.5).

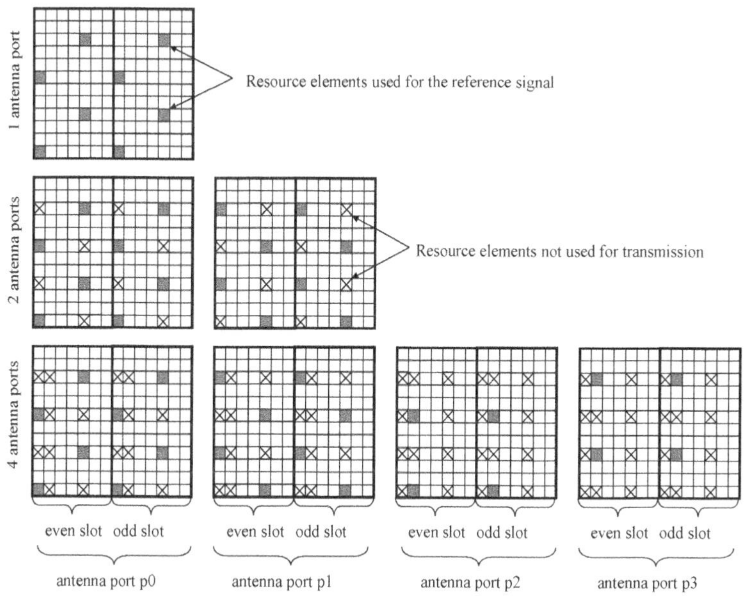

Figure 6.5. *Mapping of the Cell-Specific RS physical signal for a normal cyclic prefix*

The resource elements used for the antenna port p2 (p3, respectively) are situated at the same level, in the frequency domain, of those used for the first symbol in the case of the antenna port p0 (p1, respectively).

For an extended cyclic prefix and for antenna ports p0 and p1, the mapping takes place on the first and fourth OFDM symbol of each time slot (Figure 6.6).

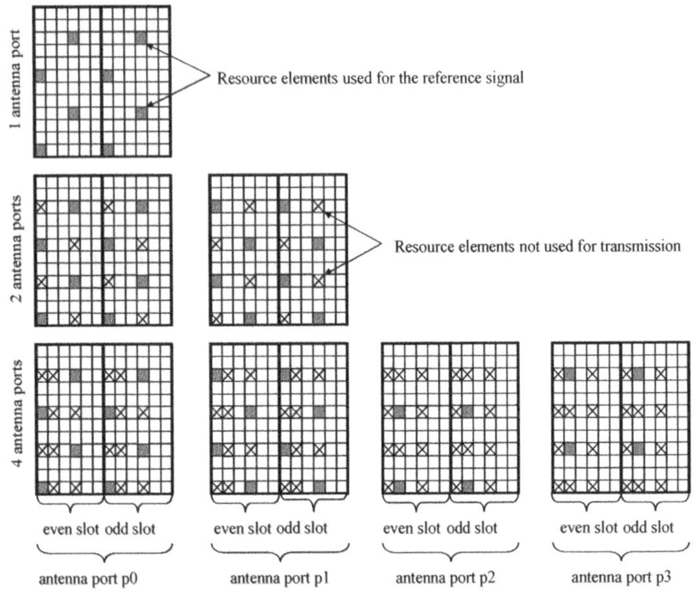

Figure 6.6. *Mapping of the Cell-Specific RS physical signal for an extended cyclic prefix*

For an extended cyclic prefix and for antenna ports p2 and p3, the mapping remains unchanged as compared to that used for the normal cyclic prefix.

For the case of the antenna ports p0 and p1, the resource elements allocated to the Cell-Specific RS physical signal in the time domain, make it possible to trace the temporary variations of the radio channel, and are therefore adapted to a high-mobility environment.

For the case of the antenna ports p2 and p3, the resource elements allocated to the Cell-Specific RS physical signal in the time domain are two times lower and are therefore more adapted to a low-mobility environment.

Two adjacent cells must map differently the Cell-Specific RS physical signal, in the frequency domain, in order to avoid inter-cell interference when the cells are synchronized; this is the case when three cells are controlled on the same site (Figure 6.7).

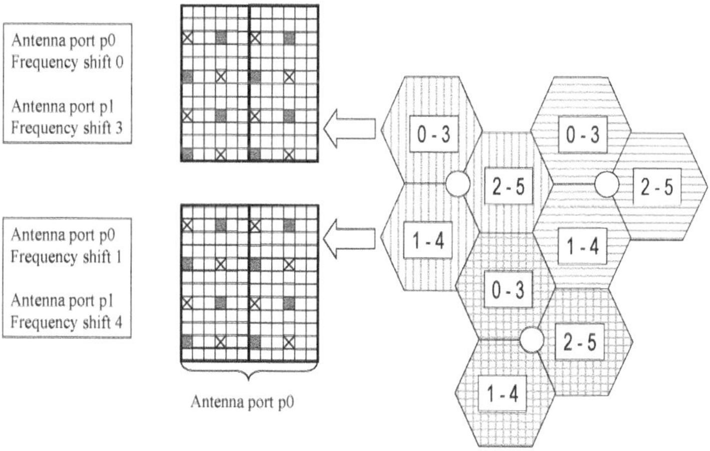

Figure 6.7. *Frequency planning of the Cell-Specific RS physical signal*

6.4. MBSFN RS physical signal

The MBMS Single Frequency Network Reference Signal (MBSFN RS) is only transmitted over the Physical Multicast Channel (PMCH) to perform the coherent demodulation.

The MBSFN RS physical signal is associated with the antenna port p4.

The MBSFN RS physical signal is only defined in the case of an extended cyclic prefix.

The MBSFN RS physical signal is defined for a 15 kHz or 7.5 kHz spacing between the subcarriers.

6.4.1. *Sequence generation*

The MBSFN RS physical signal is obtained from the following sequence:

$$r_{l,n_s}(m) = \frac{1}{\sqrt{2}}\left(1 - 2 \cdot c(2m)\right) + j\frac{1}{\sqrt{2}}\left(1 - 2 \cdot c(2m+1)\right),$$

$$m = 0,1,...,6N_{\mathrm{RB}}^{\max,\mathrm{DL}} - 1$$

$N_{\text{RB}}^{\max,\,\text{DL}}$ is expressed as the number of the Resource Blocks (RB) in the frequency domain for the entire bandwidth of the radio channel.

l corresponds to the number of the Orthogonal Frequency Division Multiplexing (OFDM) symbol of the time slot.

n_s corresponds to the time slot number of the time frame.

$c(m)$ corresponds to a pseudo-random sequence that is 31-bit long.

At the start of every OFDM symbol, the initial value c_{init} of the pseudo-random sequence is obtained with the following formula:

$$c_{\text{init}} = 2^9 \cdot \left(7 \cdot \left(n_s + 1\right) + l + 1\right) \cdot \left(2 \cdot N_{\text{ID}}^{\text{MBSFN}} + 1\right) + N_{\text{ID}}^{\text{MBSFN}}$$

$N_{\text{ID}}^{\text{MBSFN}}$ corresponds to the identity of the MBSFN area.

6.4.2. *Mapping on the resource elements*

Figure 6.8 describes the mapping on the Resource Elements (RE) for the case of a 15 kHz spacing between subcarriers.

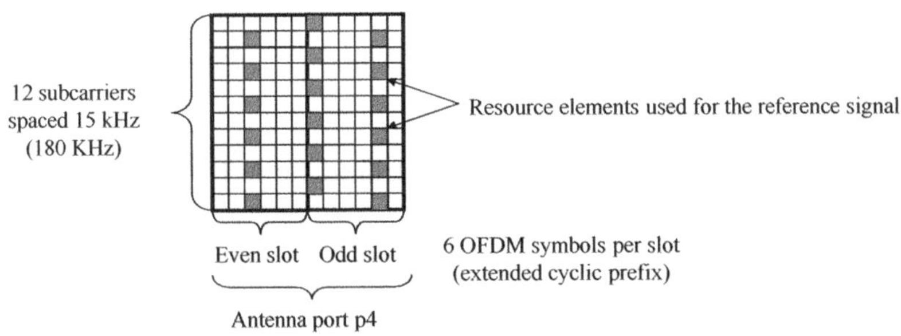

Figure 6.8. *Mapping of the MBSFN RS physical signal for a 15 kHz spacing between the subcarriers*

The MBSFN RS physical signal is mapped, in the time domain, on the resource elements located in the third OFDM symbol of the even time slot, and on the first and fifth OFDM symbols of the odd time slot.

The MBSFN RS physical signal is mapped, in the frequency domain, on the following resource elements:

– even-numbered resource elements for the third OFDM symbol of the even time slot and for the fifth OFDM symbol of the odd time slot;

– odd-numbered resource elements for the first OFDM symbol of the odd time slot.

Figure 6.9 describes the mapping on the resource elements in the case of a 7.5 kHz spacing between the subcarriers.

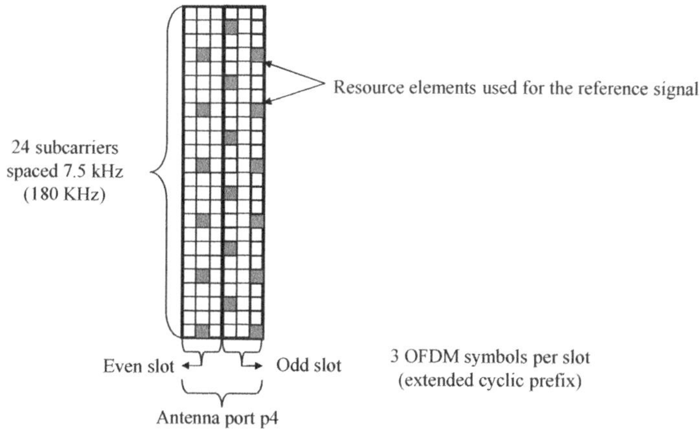

Figure 6.9. *Mapping of the MBSFN RS physical signal for a 7.5 kHz spacing between the subcarriers*

The MBSFN RS physical signal is mapped, in the time domain, on the resource elements located in the second OFDM symbol of the even time slot, and on the first and the third OFDM symbols of the odd time slot.

The MBSFN RS physical signal is mapped, in the frequency domain, on the following resource elements:

– resource elements whose rank is a multiple of 4 for the second OFDM symbol of the even time slot and for the third OFDM symbol of the odd time slot;

– resource elements whose rank has a two-subcarrier shift for the first OFDM symbol of the odd time slot.

6.5. UE-Specific RS physical signal

The UE-Specific Reference Signal (RS) is used for the beamforming mechanism and allows for the demodulation of the Physical Downlink Shared Channel (PDSCH).

UE-Specific RS physical signal is solely transmitted over the PDSCH physical channel and uses the following antenna ports:

– $p=5$ in the case of Release 8;

– $p=7$, $p=8$ in the case of Release 9;

– $p=7,8,...,\upsilon+6$, in the case of Release 10, where υ denotes the number of spatial layers.

Release 8 allows the user configuration employing a single spatial layer.

Release 9 allows the following configurations:

– one user that employs two spatial layers, and one beamforming associated with 2×2 Multiple Input Multiple Output (2×2 MIMO) spatial multiplexing;

– two users that employ two spatial layers; the two-user multiplexing is performed by an identity code;

– four users that employ one spatial layer, the multiplexing of the four spatial layers being obtained from the Orthogonal Covering Code (OCC) and from an identity code.

Release 10 allows the configuration with a user that employs eight spatial layers, and one beamforming associated with a spatial multiplexing using 8×8 MIMO mechanism.

The UE-Specific RS physical signal coexists with the Cell-Specific RS physical signal, both reference signals being present in the resource blocks allocated to the mobile.

6.5.1. *Sequence generation*

For antenna port p5, the UE-Specific RS physical signal is obtained from the following sequence:

$$r_{n_s}(m)=\frac{1}{\sqrt{2}}\left(1-2\cdot c(2m)\right)+j\frac{1}{\sqrt{2}}\left(1-2\cdot c(2m+1)\right)$$

$$m = 0,1,...,12N_{\mathrm{RB}}^{\mathrm{PDSCH}} - 1$$

$N_{\mathrm{RB}}^{\mathrm{PDSCH}}$ is expressed as the number of Resource Blocks (RB) allocated to the mobile in the PDSCH channel.

n_{s} corresponds to the time slot number of the time frame.

$c(m)$ corresponds to a pseudo-random sequence that is 31-bit long.

At the start of each subframe, the initial value c_{init} of the pseudo-random sequence is calculated as follows:

$$c_{\mathrm{init}} = \left(\lfloor n_{\mathrm{s}}/2 \rfloor + 1\right) \cdot \left(2N_{\mathrm{ID}}^{\mathrm{cell}} + 1\right) \cdot 2^{16} + n_{\mathrm{RNTI}}$$

$N_{\mathrm{ID}}^{\mathrm{cell}}$ corresponds to the Physical-layer Cell Identity (PCI).

n_{RNTI} corresponds to the Radio Network Temporary Identifier (RNTI), allocated to the mobile during the connection.

For the antenna ports $p \in \{7,8,...,\upsilon+6\}$, the UE-Specific RS physical signal is obtained as follows:

$$r(m) = \frac{1}{\sqrt{2}}(1 - 2 \cdot c(2m)) + j\frac{1}{\sqrt{2}}(1 - 2 \cdot c(2m+1))$$

$$m = \begin{cases} 0,1,...,12N_{\mathrm{RB}}^{\mathrm{max,DL}} - 1 & \text{for a normal cyclic prefix} \\ 0,1,...,16N_{\mathrm{RB}}^{\mathrm{max,DL}} - 1 & \text{for an extended cyclic prefix} \end{cases}$$

$N_{\mathrm{RB}}^{\mathrm{max,\,DL}}$ is expressed as the number of resource blocks in the frequency domain for the entire bandwidth of the radio channel.

At the start of each subframe, the initial value c_{init} of the pseudo-random sequence is calculated as follows:

$$c_{\mathrm{init}} = \left(\lfloor n_{\mathrm{s}}/2 \rfloor + 1\right) \cdot \left(2N_{\mathrm{ID}}^{\mathrm{cell}} + 1\right) \cdot 2^{16} + n_{\mathrm{SCID}}$$

n_{SCID} is the identity of the scrambling code. It takes the ZERO value, except where the value is indicated in the Downlink Control Information (DCI), for the formats 2B and 2C.

6.5.2. *Mapping on the resource elements*

For the antenna port p5, in the case of a normal prefix, the UE-Specific RS physical signal is mapped, in the time domain, on the Resource Elements (RE) located in the following Orthogonal Frequency Division Multiplexing (OFDM) symbols (Figure 6.10):

– in the fourth and seventh OFDM symbols of the even time slot;

– in the third and sixth OFDM symbols of the odd time slot.

The UE-Specific RS physical signal is mapped, in the frequency domain, on the following resource elements:

– resource elements whose rank is a multiple of 4 for the fourth OFDM symbol of the even time slot and for the third OFDM symbol of the odd time slot;

– resource elements whose rank has a two-subcarrier shift for the seventh OFDM symbol of the even time slot and for the sixth OFDM symbol of the odd time slot.

The UE-Specific RS physical signal can take four positions in the frequency domain; each position depends on the value of the cell physical identity.

For the antenna port p5, in the case of an extended prefix, the UE-Specific RS physical signal is mapped, in the time domain, on resource elements situated in the following OFDM symbols (Figure 6.10):

– in the fifth OFDM symbol for the even time slot;

– in the second and fifth OFDM symbols for the odd time slot.

The UE-Specific RS physical signal is mapped, in the frequency domain, on the following resource elements:

– resource elements whose rank is a multiple of 3 for the fifth OFDM symbol of the even time slot and of the odd time slot;

– resource elements whose rank has a two-subcarrier shift for the second OFDM symbol of the odd time slot.

The UE-Specific RS physical signal can take three positions in the frequency domain; each position depends on the value $N_{\text{ID}}^{\text{cell}}$ of the cell physical identity.

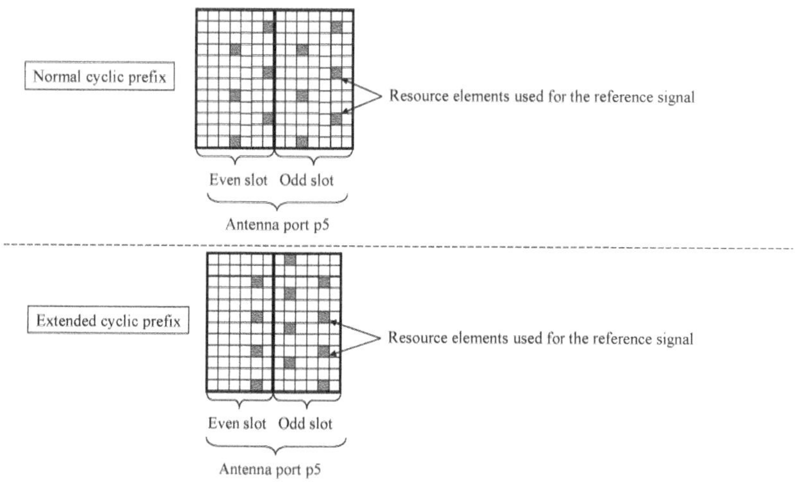

Figure 6.10. *Mapping of the UE-Specific RS physical signal for antenna port p5*

For the case of the antenna ports p7 to p14, the mapping depends on several factors:

– the type of cyclic prefix, normal or extended. In the case of the extended prefix, ports p7 and p8 are the only ones to be used;

– the configuration of special subframes for the Time Division Duplex (TDD) mode;

– the antenna port number.

For the antenna ports p7, p8, p11 and p 13, in the case of a normal prefix, the mapping takes place on the sixth and seventh OFDM symbol of each time slot (Figure 6.11).

For the antenna ports p9, p10, p12 and p14, in the case of a normal prefix, the mapping takes place on the sixth and seventh OFDM symbol of each time slot, with a one-subcarrier shift in the frequency domain (Figure 6.11).

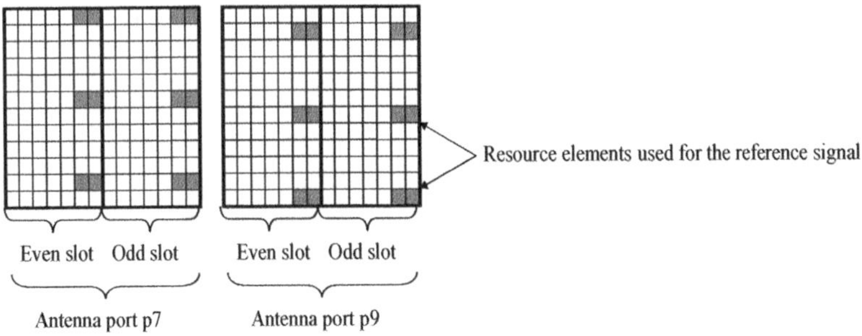

Figure 6.11. *Mapping of the UE-Specific RS physical signal for antenna ports p7 and p9, in the case of a normal prefix*

The resource elements allocated to the UE-Specific RS physical signal are identical for four antenna ports (7, 8, 11, 13 on the one hand, and, 9, 10, 12, 14 on the other hand). The sequence orthogonality is obtained with an OCC code (Table 6.2).

Antenna Port	OCC code
7	$[+1 \quad +1 \quad +1 \quad +1]$
8	$[+1 \quad -1 \quad +1 \quad -1]$
9	$[+1 \quad +1 \quad +1 \quad +1]$
10	$[+1 \quad -1 \quad +1 \quad -1]$
11	$[+1 \quad +1 \quad -1 \quad -1]$
12	$[-1 \quad -1 \quad +1 \quad +1]$
13	$[+1 \quad -1 \quad -1 \quad +1]$
14	$[-1 \quad +1 \quad +1 \quad -1]$

Table 6.2. *Scrambling codes for the case of a normal prefix*

Figure 6.12 describes the mapping on the resource elements of special subframes of the TDD mode, for antenna ports 7 and 9, in the case of a normal prefix.

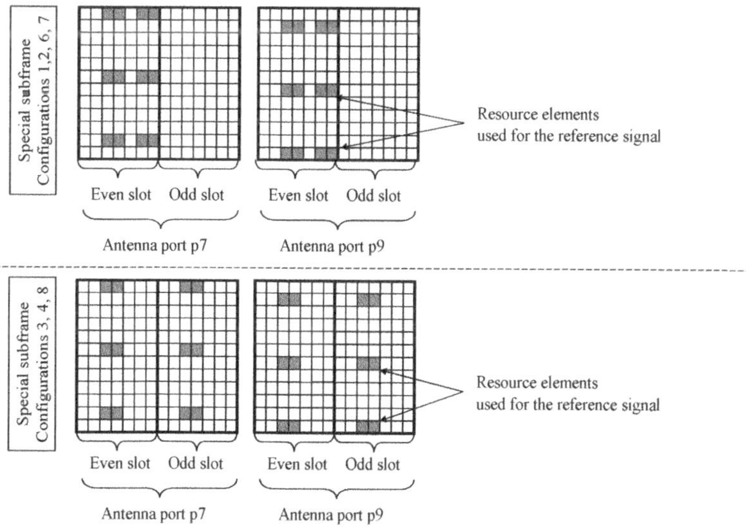

Figure 6.12. *Mapping of the UE-Specific RS physical signal applied to the special subframes of the TDD mode for antenna ports p7 and p9, in the case of a normal prefix*

In the case of an extended prefix, the resource elements allocated to the UE-Specific RS physical signal are identical for both antenna ports p7 and p8. The sequence orthogonality is obtained with an OCC code (Table 6.3).

Antenna Port	OCC code
7	$[+1 \quad +1]$
8	$[-1 \quad +1]$

Table 6.3. *Scrambling codes for the case of an extended prefix*

Figure 6.13 describes the mapping on the resource elements for antenna ports 7, in the case of an extended prefix.

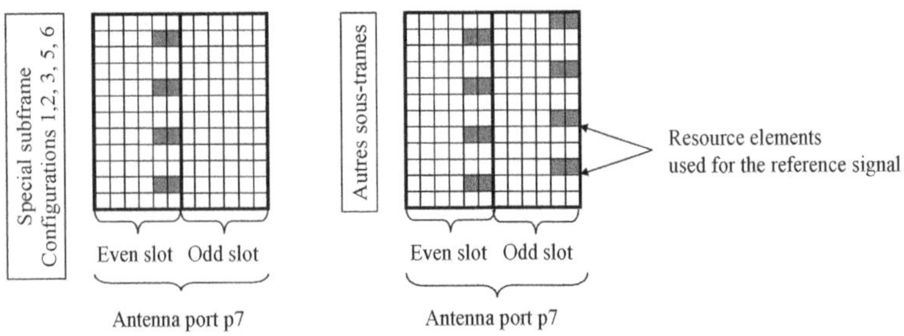

Figure 6.13. *Mapping of the UE-Specific RS physical signal for the antenna port p7, in the case of an extended prefix*

6.6. PRS physical signal

The Positioning Reference Signal (PRS) was introduced in Release 9.

The PRS physical signal is used by the mobile for the implementation of the Observed Time Difference of Arrival (OTDOA) function.

The function is based on the difference in reception time of the PRS physical signal, as measured by the mobile, in comparison with a reference cell.

The location of the mobile is obtained from three measures performed on three geographically-dispersed cells.

The PRS physical signal is associated with the antenna port p6.

The PRS physical signal must not be mapped on the resource elements used for the Primary Synchronization Signal (PSS), for the Secondary Synchronization Signal (SSS) and for the Physical Broadcast Channel (PBCH).

The PRS physical signal must not be mapped to the resource elements used by the special subframes of the Time Division Duplex (TDD) mode.

The PRS physical signal is defined by a 15 kHz spacing between the subcarriers.

The PRS physical signal coexists with the Cell-Specific RS physical signal, both reference signals being present in the resource blocks allocated to the mobile.

6.6.1. *Sequence generation*

The PRS physical signal is obtained with the following sequence:

$$r_{l,n_s}(m) = \frac{1}{\sqrt{2}}\left(1-2\cdot c(2m)\right)+j\frac{1}{\sqrt{2}}\left(1-2\cdot c(2m+1)\right)$$

$$m = 0,1,\ldots,2N_{\mathrm{RB}}^{\mathrm{max,DL}}-1$$

$N_{\mathrm{RB}}^{\mathrm{max,\,DL}}$ is expressed as the number of resource blocks in the frequency domain, for the entire bandwidth of the radio channel.

l corresponds to the OFDM symbol number of the time slot.

n_{s} corresponds to the time slot number of the time frame.

$c(m)$ corresponds to a pseudo-random sequence that is 31 bits long.

At the start of each OFDM symbol, the initial value c_{init} of the pseudo-random sequence is obtained with the following formula:

$$c_{\mathrm{init}} = 2^{10}\cdot\left(7\cdot\left(n_{\mathrm{s}}+1\right)+l+1\right)\cdot\left(2\cdot N_{\mathrm{ID}}^{\mathrm{cell}}+1\right)+2\cdot N_{\mathrm{ID}}^{\mathrm{cell}}+N_{\mathrm{CP}}$$

$N_{\mathrm{ID}}^{\mathrm{cell}}$ corresponds to the Physical-layer Cell Identity (PCI).

N_{CP} takes a value equal to ONE for a normal cyclic prefix and to ZERO for an extended cyclic prefix.

6.6.2. *Mapping on the resource elements*

The mapping of the PRS physical signal depends on the following factors:

– the type of cyclic prefix: normal or extended;

– the configuration of the antenna ports (2 antenna ports p0 and p1, or 4 antenna ports p0, p1, p2 and p3) used for the Cell-Specific RS physical signal.

The resource elements used for the PRS physical signal have a spacing of 6 subcarriers in the frequency domain.

The PRS physical signal can take six positions in the frequency domain; each position depends on the value $N_{\mathrm{ID}}^{\mathrm{cell}}$ of the physical identity of the cell.

Figure 6.14 describes the mapping on the resource elements in the case of a normal cyclic prefix.

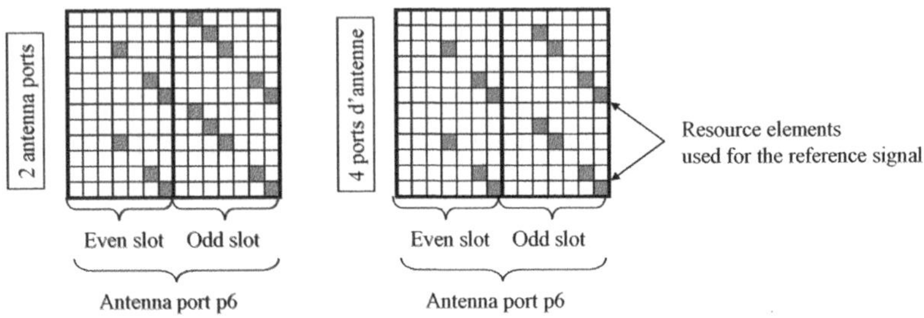

Figure 6.14. *Mapping of the PRS physical signal for a normal cyclic prefix*

Figure 6.15 describes the mapping on the resource elements in the case of an extended cyclic prefix.

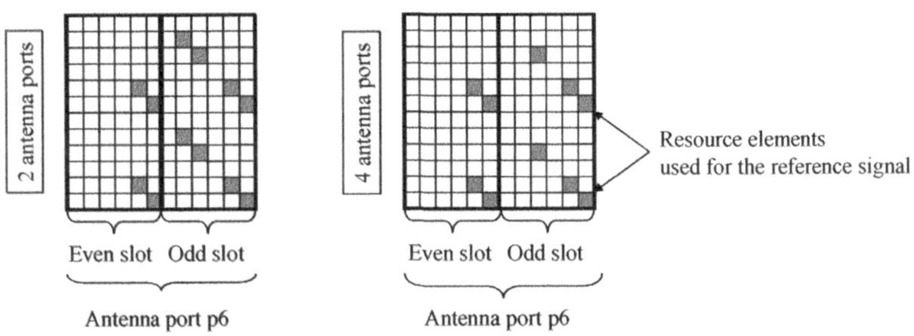

Figure 6.15. *Mapping of the PRS physical signal for an extended cyclic prefix*

6.6.3. *Configuration of the subframes*

The configuration of the subframes containing the PRS physical signal is defined in the LTE Positioning Protocol (LPP) message.

The bandwidth $N_{\mathrm{RB}}^{\mathrm{PRS}}$ allocated to the PRS physical signal is expressed as the number of resource blocks in the frequency domain.

The configuration index I_{PRS} determines the periodicity T_{PRS} for transmission of the PRS physical signal and for the shift Δ_{PRS} (Table 6.4).

Configuration index I_{PRS}	Periodicity T_{PRS}	Shift Δ_{PRS}
0–159	160	I_{PRS}
160–479	320	$I_{\text{PRS}}-160$
480–1119	640	$I_{\text{PRS}}-480$
1120–2399	1280	$I_{\text{PRS}}-1120$
2400–4095	Reserved	

Table 6.4. *Configuration parameters for the mapping of the PRS physical signal*

The number of subframes in a group, in which the reference signal is present, can take the value 1, 2, 4 or 6.

The first subframe number in the group is obtained with the following formula:

$$\left(10\times n_{\text{f}}+\lfloor n_{\text{s}}/2\rfloor-\Delta_{\text{PRS}}\right)\bmod T_{\text{PRS}}=0$$

n_{f} corresponds to the time frame number;

n_{s} corresponds to the time slot number of the time frame.

6.7. CSI RS physical signal

The Channel State Information Reference Signal (CSI RS) was introduced in Release 10.

The CSI RS physical signal helps to improve the measurement of the received signal and of the interference level as compared to that supplied by means of Cell-Specific RS physical signal.

The power of CSI RS physical signal is either transmitted to determine the level of the received signal or deleted to measure the interference level.

The CSI RS physical signal occupies new resource elements in the Physical Downlink Shared Channel (PDSCH).

The CSI RS physical signal can be associated with one antenna port p15, two antenna ports p15 and p16, four antenna ports p15 to p18 or eight antenna ports p15 to p22.

The CSI RS physical signal is defined for 15 kHz spacing between subcarriers.

The CSI RS physical signal is not transmitted in the following subframes:

– special subframes for type-2 frames;

– subframes containing the System Information Block 1 (SIB1);

– subframes containing the paging messages.

6.7.1. *Sequence generation*

The CSI RS physical signal is obtained with the following sequence:

$$r_{l,n_s}(m) = \frac{1}{\sqrt{2}}\left(1-2\cdot c(2m)\right) + j\frac{1}{\sqrt{2}}\left(1-2\cdot c(2m+1)\right)$$

$$m = 0,1,\ldots,N_{\mathrm{RB}}^{\mathrm{max,DL}} - 1$$

$N_{\mathrm{RB}}^{\mathrm{max,\,DL}}$ is expressed as the number of resource blocks in the frequency domain, for the entire bandwidth of the radio channel.

l corresponds to the Orthogonal Frequency Division Multiplexing (OFDM) symbol number of the time slot.

n_s corresponds to the time slot number of the time frame.

$c(m)$ corresponds to a pseudo-random sequence that is 31-bit long.

At the start of each OFDM symbol, the initial value c_{init} of the pseudo-random sequence is obtained with the following formula:

$$c_{\mathrm{init}} = 2^{10}\cdot\left(7\cdot\left(n_s+1\right)+l+1\right)\cdot\left(2\cdot N_{\mathrm{ID}}^{\mathrm{cell}}+1\right)+2\cdot N_{\mathrm{ID}}^{\mathrm{cell}}+N_{\mathrm{CP}}$$

$N_{\mathrm{ID}}^{\mathrm{cell}}$ corresponds to the Physical-layer Cell Identity (PCI).

N_{CP} takes a value equal to ONE for a normal cyclic prefix and to ZERO for an extended cyclic prefix.

6.7.2. *Mapping on the resource elements*

The mapping on the resource elements is defined in the RRC messages *ConnectionSetup*, *ConnectionReconfiguration*, *ConnectionRe-establishment*:

– the number of the antenna ports takes a value equal to 1, 2, 4 or 8;

– the mapping configuration index takes a value ranging between 0 and 31 in the case of a normal cyclic prefix, or between 0 and 27 in the case of an extended cyclic prefix.

For instance, Figures 6.16 and 6.17 describe the mapping on resource elements for the configuration index 0, in the case of a normal and extended cyclic prefix.

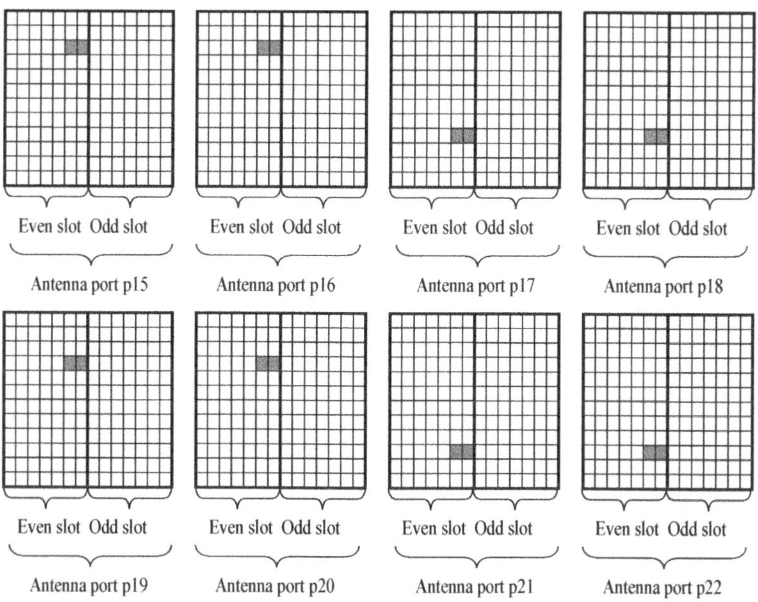

Figure 6.16. *Mapping of the CSI RS physical signal: configuration 0, normal cyclic prefix*

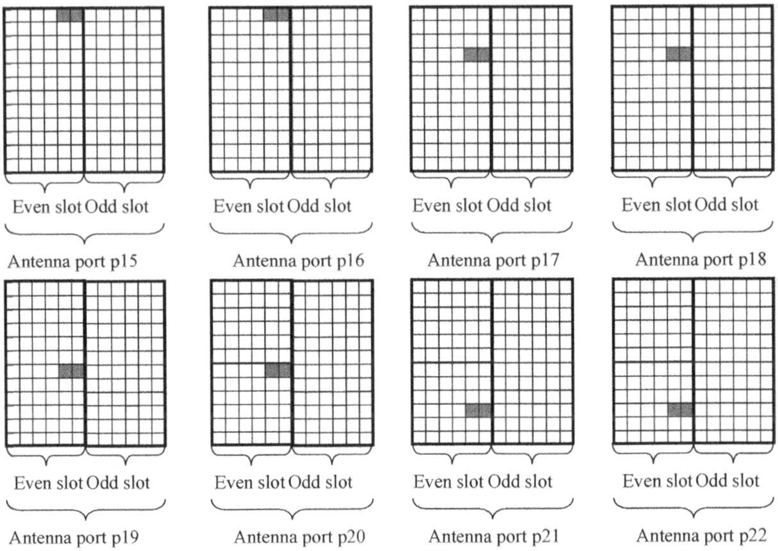

Figure 6.17. *Mapping of the CSI RS physical signal: configuration 0, extended cyclic prefix*

6.7.3. *Configuration of the subframes*

The configuration of the subframes containing the CSI RS physical signal, with or without transmit power, is defined in the RRC messages *ConnectionSetup*, *ConnectionReconfiguration*, *ConnectionRe-establishment*.

The configuration index I_{CSI-RS} determines the periodicity T_{CSI-RS} of transmission of the CSI RS physical signal and the shift Δ_{CSI-RS} (Table 6.5).

Configuration index I_{CSI-RS}	Periodicity T_{CSI-RS}	Shift Δ_{CSI-RS}
0–4	5	I_{CSI-RS}
5–14	10	$I_{CSI-RS} - 5$
15–34	20	$I_{CSI-RS} - 15$
35–74	40	$I_{CSI-RS} - 35$

Table 6.5. *Configuration parameters for the mapping of the CSI RS physical signal*

The number of the subframe that contains the CSI RS physical signal is obtained with the following formula:

$$\left(10\times n_{\mathrm{f}}+\lfloor n_{\mathrm{s}}/2\rfloor-\Delta_{\mathrm{CSI-RS}}\right)\bmod T_{\mathrm{CSI-RS}}=0$$

n_{f} corresponds to the time frame number.

7

Downlink Physical Channels

7.1. PBCH physical channel

The Physical Broadcast Channel (PBCH) transmits the BCH transport channel that contains the system information corresponding to the Master Information Block (MIB) message.

The processing associated with the PBCH physical channel is summarized in Figure 7.1.

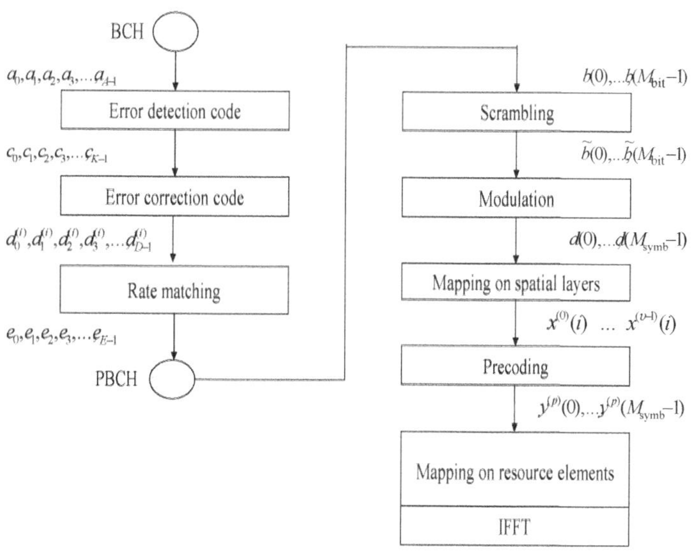

Figure 7.1. *Processing associated with the PBCH physical channel*

7.1.1. *Error detection code*

The error detection is obtained with a Cyclic Redundancy Check (CRC) code.

The CRC structure is the remainder when the transport block $a_0, a_1, a_2, a_3, ..., a_{A-1}$ (A = 24 bits) is divided by a generator polynomial, whose remainder $p_0, p_1, p_2, p_3, ..., p_{L-1}$ (L=16) constitutes the cyclic redundancy bits.

The 16 cyclic redundancy bits $p_0, p_1, p_2, p_3, ..., p_{L-1}$ are scrambled by the sequence $x_{ant,0}, x_{ant,1}, ..., x_{ant,15}$, then are concatenated to the transport block to constitute the sequence $c_0, c_1, c_2, c_3, ..., c_{K-1}$.

The sequence $x_{ant,0}, x_{ant,1}, ..., x_{ant,15}$ determines the number of antenna ports (Table 7.1).

Number of antenna ports	Sequence $x_{ant,0}, x_{ant,1}, ..., x_{ant,15}$
1	<0, 0, 0, 0, 0, 0, 0, 0, 0, 0, 0, 0, 0, 0, 0, 0>
2	<1, 1, 1, 1, 1, 1, 1, 1, 1, 1, 1, 1, 1, 1, 1, 1>
4	<0, 1, 0, 1, 0, 1, 0, 1, 0, 1, 0, 1, 0, 1, 0, 1>

Table 7.1. *Mask of the cyclic redundancy sequence*

7.1.2. *Error correction code*

The error correction code is a convolutional code that produces three sequences $d_0^{(i)}, d_1^{(i)}, d_2^{(i)}, d_3^{(i)}, ..., d_{D-1}^{(i)}$, where $i = 0$, 1 and 2 and D corresponds to the number of bits of the sequence.

7.1.3. *Rate matching*

The three sequences $d_0^{(i)}, d_1^{(i)}, d_2^{(i)}, d_3^{(i)}, ..., d_{D-1}^{(i)}$ derived from the convolutional code are repeated to reach a 1920-bit sequence in the case of a normal prefix or a 1728-bit sequence in the case of an extended prefix.

The output sequence is noted $e_0, e_1, e_2, e_3, ..., e_{E-1}$, where E corresponds to the number of bits of the sequence.

7.1.4. *Scrambling*

The bit sequence $b(0),...,b(M_{\text{bit}}-1)$, M_{bit} that identifies the number of bits transmitted to the PBCH physical channel, is scrambled by the sequence $c(i)$.

This results is the bit sequence $\tilde{b}(0),...,\tilde{b}(M_{\text{bit}}-1)$ with $\tilde{b}(i) = \left(b(i) + c(i)\right) \bmod 2$.

The scrambling sequence is built with a pseudo-random sequence that uses a 31-bit polynomial.

The initial value c_{init} of the scrambling sequence equals $N_{\text{ID}}^{\text{cell}}$ that corresponds to the Physical-layer Cell Identity (PCI) retrieved when the time synchronization of the mobile was performed.

7.1.5. *Modulation*

The symbol sequence $d(0),...,d(M_{\text{symb}}-1)$, corresponding to a Quadrature Phase-Shift Keying (QPSK) modulation, is obtained from the bit sequence $\tilde{b}(0),...,\tilde{b}(M_{\text{bit}}-1)$, a symbol corresponding to two bits.

The number of symbols M_{symb} equals 960 in the case of a normal cyclic prefix and 864 in the case of an extended cyclic prefix.

7.1.6. *Mapping on the spatial layers*

The set of symbols $d\ (0),...,d\ (M_{\text{symb}}-1)$ is mapped to the set of symbols $x^{(0)}(i)$ ... $x^{(\upsilon-1)}(i)$, with $i = 0,1,...,M_{\text{symb}}^{\text{layer}}-1$, where υ corresponds to the number of spatial layers and $M_{\text{symb}}^{\text{layer}}$ corresponds to the number of symbols per spatial layer.

The number of spatial layers corresponds to the number of antenna ports.

When one antenna port (p0) is used, the following relation is obtained:

$$x^{(0)}(i) = d^{(0)}(i)$$

When two antenna ports (p0 and p1) are used, the following relations are obtained:

$$x^{(0)}(i) = d^{(0)}(2i)$$
$$x^{(1)}(i) = d^{(0)}(2i+1)$$

$$M_{\text{symb}}^{\text{layer}} = M_{\text{symb}}/2$$

When four antenna ports (p0 to p3) are used, the following relations are obtained:

$$x^{(0)}(i) = d^{(0)}(4i)$$
$$x^{(1)}(i) = d^{(0)}(4i+1)$$
$$x^{(2)}(i) = d^{(0)}(4i+2)$$
$$x^{(3)}(i) = d^{(0)}(4i+3)$$

$$M_{\text{symb}}^{\text{layer}} = M_{\text{symb}}/4$$

7.1.7. *Precoding*

The precoding is applied to the set of symbols $x^{(0)}(i)$ ... $x^{(\upsilon-1)}(i)$ to generate the sequences $y^{(p)}(0),...,y^{(p)}(M_{\text{symb}}-1)$, where p corresponds to the antenna port number.

When one single antenna port is used (p0), no precoding is performed.

When two antenna ports are used (p0 and p1), the PBCH physical channel is transmitted over Space Frequency Block Coding (SFBC), a form of transmit diversity.

The symbols are transmitted using the following precoding:

– antenna port (0) $y^{(0)}(2i) = x^{(0)}(i)$ $y^{(0)}(2i+1) = x^{(1)}(i)$;

– antenna port (1) $y^{(1)}(2i) = -(x^{(1)}(i))^*$ $y^{(1)}(2i+1) = (x^{(0)}(i))^*$.

The acronym (*) represents the complex conjugate.

When four antenna ports (p0 to p3) are used, the PBCH physical channel is transmitted over SFBC/FSTD (Frequency Shift Transmit Diversity).

The symbols are transmitted using the following precoding:

– antenna port (0) $y^{(0)}(4i) = x^{(0)}(i)$ $y^{(0)}(4i+1) = x^{(1)}(i)$;

– antenna port (1) $y^{(1)}(4i+2) = x^{(2)}(i)$ $y^{(1)}(4i+3) = x^{(3)}(i)$;

– antenna port (2) $y^{(2)}(4i) = -(x^{(1)}(i))^*$ $y^{(2)}(4i+1) = (x^{(0)}(i))^*$;

– antenna port (3) $y^{(3)}(4i+2) = -(x^{(3)}(i))^*$ $y^{(3)}(4i+3) = (x^{(2)}(i))^*$.

7.1.8. *Mapping on the resource elements*

The sequence of symbols $y^{(p)}(0),...,y^{(p)}(M_{\text{symb}}-1)$ can be considered as 4 blocks comprising 240 symbols in the case of the normal cyclic prefix, or 216 symbols in the case of the extended cyclic prefix.

The sequence of symbols $y^{(p)}(0),...,y^{(p)}(M_{\text{symb}}-1)$ for each antenna port is transmitted in four consecutive frames, where the number of the first frame must be a multiple of 4.

In the frequency domain, the sequence of symbols is mapped to 72 subcarriers located on both sides of the subcarrier that corresponds to the central frequency of the radio channel.

In the time domain, one block with 240 or 216 symbols is mapped to the first four Orthogonal Frequency Division Multiplexing (OFDM) symbols of the second time slot (slot 1) of subframe 0.

The resource elements used for the Cell-Specific Reference Signal (RS), that comprise all those allocated to the four antenna ports p0 to p3, are not used for the PBCH physical channel.

The mapping of the PBCH physical channel, for a normal cyclic prefix and for the Frequency Division Duplex (FDD) mode, is outlined in Figure 7.2.

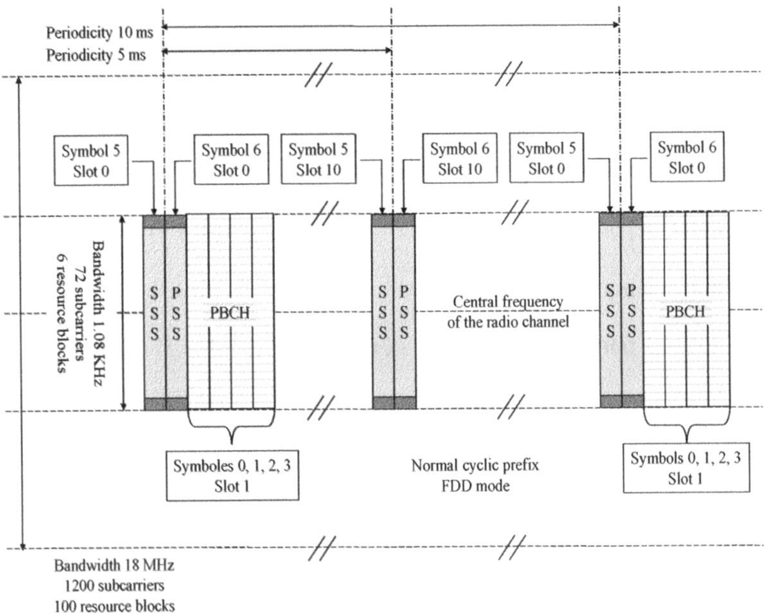

Figure 7.2. *Mapping of the PBCH physical channel*

7.2. PCFICH physical channel

The Control Format Indicator (CFI) defines, for each subframe, the number of Orthogonal Frequency Division Multiplexing (OFDM) symbols allocated to the Physical Downlink Control Channel (PDCCH).

The CFI information is transmitted over the Physical Control Format Indicator Channel (PCFICH).

The processing associated with the PCFICH physical channel is summarized in Figure 7.3.

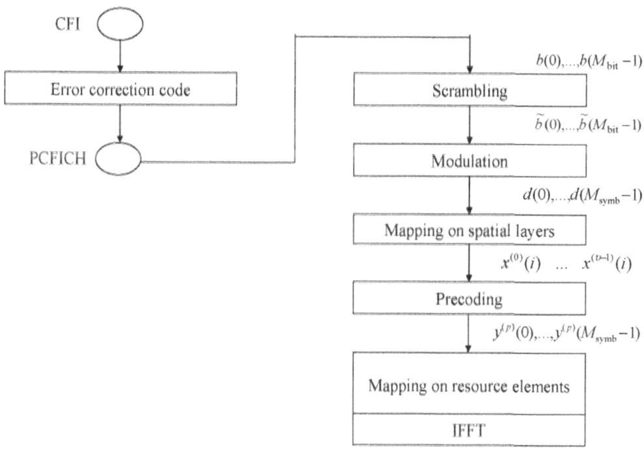

Figure 7.3. *Processing associated with the PCFICH physical channel*

7.2.1. *CFI information*

The CFI information takes a value equal to 1, 2 or 3.

If the bandwidth of the radio channel is greater than 1.4-MHz, the number of the OFDM symbols (1, 2, 3) is equal to the value of the CFI.

If the bandwidth of the radio channel equals 1.4-MHz, the number of the OFDM symbols (2, 3, 4) is equal to the CFI value plus 1.

7.2.2. *Error correction code*

The code per block generates a 32-bit sequence $b(0),...,b(31)$ by means of two bits that encode the value of the CFI information (Table 7.2).

CFI value	Sequence $b(0),....,b(31)$
1	0,1,1,0,1,1,0,1,1,0,1,1,0,1,1,0,1,1,0,1,1,0,1,1,0,1,1,0,1,1,0,1
2	1,0,1,1,0,1,1,0,1,1,0,1,1,0,1,1,0,1,1,0,1,1,0,1,1,0,1,1,0,1,1,0
3	1,1,0,1,1,0,1,1,0,1,1,0,1,1,0,1,1,0,1,1,0,1,1,0,1,1,0,1,1,0,1,1
4	0,0,0,0,0,0,0,0,0,0,0,0,0,0,0,0,0,0,0,0,0,0,0,0,0,0,0,0,0,0,0,0

Table 7.2. *Coding of the CFI information*

7.2.3. *Scrambling*

The bit sequence $b(0),...,b(31)$ is scrambled with a pseudo-random sequence $c(i)$ that uses a 31-bit polynomial, to produce the bit sequence $\tilde{b}(0),...,\tilde{b}(31)$:

$$\tilde{b}(i) = \left(b(i) + c(i)\right) \bmod 2$$

The initial value c_{init} of the pseudo-random sequence is obtained with the following formula:

$$c_{\text{init}} = \left(\lfloor n_s/2 \rfloor + 1\right) \cdot \left(2N_{\text{ID}}^{\text{cell}} + 1\right) \cdot 2^9 + N_{\text{ID}}^{\text{cell}}$$

– n_s corresponds to the time slot number of the time frame;

– $N_{\text{ID}}^{\text{cell}}$ corresponds to the Physical-layer Cell Identity (PCI).

7.2.4. *Modulation*

The sequence of symbols $d(0),...,d(M_{\text{symb}} - 1)$, which corresponds to a Quadrature Phase-Shift Keying (QPSK) modulation, is generated from the bit sequence $\tilde{b}(0),...,\tilde{b}(M_{\text{bit}} - 1)$, where the number of symbols M_{symb} is equal to 16.

7.2.5. *Mapping on the spatial layers*

The set of symbols $d\,(0),...,d\,(M_{\text{symb}} - 1)$ is mapped to the set of symbols $x^{(0)}(i)$ ... $x^{(\upsilon-1)}(i)$, with $i = 0,1,...,M_{\text{symb}}^{\text{layer}} - 1$, where υ corresponds to the number of spatial layers and $M_{\text{symb}}^{\text{layer}}$ corresponds to the number of symbols per spatial layer.

The number of spatial layers corresponds to the number of antenna ports.

When an antenna port (p0) is used, the following relation is obtained:

$x^{(0)}(i) = d^{(0)}(i)$.

When two antenna ports are used (p0 and p1), the following relations are obtained:

$$x^{(0)}(i) = d^{(0)}(2i)$$
$$x^{(1)}(i) = d^{(0)}(2i+1)$$

$$M_{\text{symb}}^{\text{layer}} = M_{\text{symb}}/2 = 8$$

When four antenna ports are used (p0 to p3), the following relations are obtained:

$$x^{(0)}(i) = d^{(0)}(4i)$$
$$x^{(1)}(i) = d^{(0)}(4i+1)$$
$$x^{(2)}(i) = d^{(0)}(4i+2)$$
$$x^{(3)}(i) = d^{(0)}(4i+3)$$

$$M_{\text{symb}}^{\text{layer}} = M_{\text{symb}}/4 = 4$$

7.2.6. *Precoding*

The precoding is applied to the set of symbols $x^{(0)}(i)$ ... $x^{(\upsilon-1)}(i)$ to generate the sequences of symbols $y^{(p)}(0),...,y^{(p)}(M_{\text{symb}}-1)$, where p corresponds to the antenna port number.

When one single antenna port is used (p0), no precoding is performed.

When two antenna ports are used (p0 and p1), the PCFICH physical channel is transmitted over Space Frequency Block Coding (SFBC), a form of transmit diversity.

The symbols are transmitted using the following precoding:

– antenna port (0) $y^{(0)}(2i) = x^{(0)}(i)$ $\quad y^{(0)}(2i+1) = x^{(1)}(i)$;

– antenna Port (1) $y^{(1)}(2i) = -(x^{(1)}(i))*$ $\quad y^{(1)}(2i+1) = (x^{(0)}(i))*$.

The asterisk symbol (*) represents the complex conjugate.

When four antenna ports (p0 to p3) are used, the PCFICH physical channel is transmitted over SFBC/FSTD (Frequency Shift Transmit Diversity).

The symbols are transmitted using the following precoding:

– antenna Port (0) $y^{(0)}(4i) = x^{(0)}(i)$ $\quad y^{(0)}(4i+1) = x^{(1)}(i)$;

– antenna Port (1) $y^{(1)}(4i+2) = x^{(2)}(i)$ $\quad y^{(1)}(4i+3) = x^{(3)}(i)$;

– antenna Port (2) $y^{(2)}(4i) = -(x^{(1)}(i))*$ $\quad y^{(2)}(4i+1) = (x^{(0)}(i))*$;

– antenna Port (3) $y^{(3)}(4i+2) = -(x^{(3)}(i))*$ $\quad y^{(3)}(4i+3) = (x^{(2)}(i))*$.

7.2.7. *Mapping on the resource elements*

The sequence $y^{(p)}(0),...,y^{(p)}(M_{\text{symb}}-1)$, comprising 16 symbols, is mapped on four Resource Element Groups (REG), located in the first OFDM symbol of each subframe.

The position of each resource element group is obtained with the following relations:

– REG (0) is represented by $k = \bar{k}$;

– REG (1) is represented by $k = \bar{k} + \lfloor N_{\text{RB}}^{\text{DL}}/2 \rfloor \cdot N_{\text{sc}}^{\text{RB}}/2$;

– REG (2) is represented by $k = \bar{k} + \lfloor 2N_{\text{RB}}^{\text{DL}}/2 \rfloor \cdot N_{\text{sc}}^{\text{RB}}/2$;

– REG (3) is represented by $k = \bar{k} + \lfloor 3N_{\text{RB}}^{\text{DL}}/2 \rfloor \cdot N_{\text{sc}}^{\text{RB}}/2$.

$$\bar{k} = \left(N_{\text{sc}}^{\text{RB}}/2\right) \cdot \left(N_{\text{ID}}^{\text{cell}} \bmod 2N_{\text{RB}}^{\text{DL}}\right)$$

– k corresponds to the resource element index, in the frequency domain;

– $N_{\text{RB}}^{\text{DL}}$ corresponds to the bandwidth of the radio channel, expressed as the number of Resource Blocks (RB);

– $N_{\text{sc}}^{\text{RB}}$ corresponds to the size of the resource block in the frequency domain, expressed as the number of subcarriers.

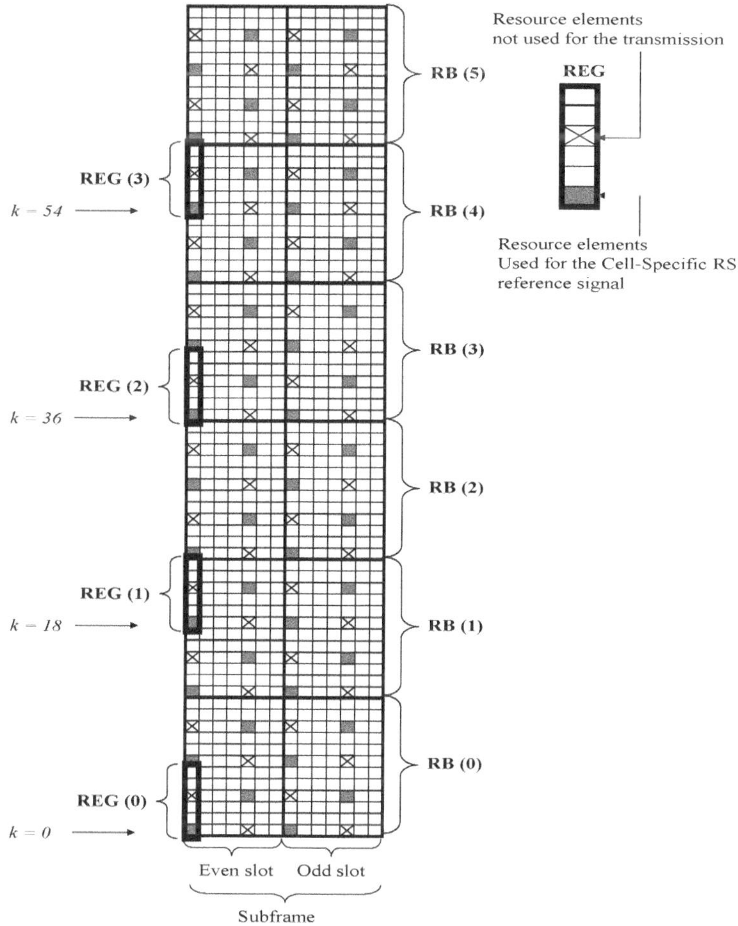

Figure 7.4. *Mapping of the PCFICH physical channel*

Figure 7.4 depicts an example of the position of the PCFICH physical channel, obtained from the following values:

– $N_{\text{RB}}^{\text{DL}} = 6$, which corresponds to a 1.4-MHz bandwidth of the radio channel;

– $N_{\text{sc}}^{\text{RB}} = 12$, which corresponds to 12 subcarriers per resource block, in the frequency domain.

$N_{\text{ID}}^{\text{cell}} = 0$

– REG (0) is represented by the index $(k = 0)$;

– REG (1) is represented by the index $(k = 18)$;

– REG (2) is represented by the index $(k = 36)$;

– REG (3) is represented by the index $(k = 54)$.

7.3. PHICH physical channel

The Hybrid Automatic Repeat reQuest (HARQ) mechanism allows one error correction per retransmission.

The evolved Node B (eNB) entity sends an HARQ Indicator (HI), transmitted in the Physical HARQ Indicator Channel (PHICH), to indicate a positive ACK or a negative NACK acknowledgment of the data received for the uplink direction in the Physical Uplink Shared Channel (PUSCH).

The processing associated with the PHICH physical channel is summarized in Figure 7.5.

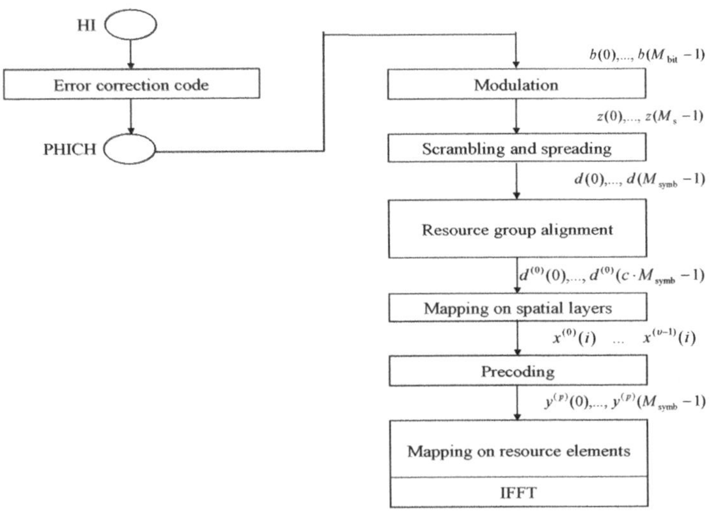

Figure 7.5. *Processing associated with the PHICH physical channel*

7.3.1. *HI information*

For Release 8 and 9, the information is encoded to one bit for the acknowledgment of a transport block. It takes a ZERO value for a NACK and a ONE value for an ACK.

For Release 10, when the Multiple Input Multiple Output (MIMO) mechanism is used, the HI information is encoded to two bits for the acknowledgment of two transport blocks.

7.3.2. *PHICH group*

Several PHICH physical channels are multiplexed in the same set of Resource Elements (RE) that represents a PHICH group.

The PHICH physical channel is identified by a group number and a sequence number in the group:

– $N_{\text{PHICH}}^{\text{group}}$ corresponds to the number of the PHICH group;

– $N_{\text{PHICH}}^{\text{seq}}$ corresponds to the number of the PHICH physical channels in the group.

For the case of a structure with a type-1 time frame that corresponds to the Frequency Division Duplex (FDD) mode, the number of PHICH groups is obtained with the following formulas:

– $N_{\text{PHICH}}^{\text{group}} = N_g \left\lceil (N_{RB}^{DL} / 8) \right\rceil$ for a normal cyclic prefix;

– $N_{\text{PHICH}}^{\text{group}} = 2 \times \left\lfloor N_g (N_{RB}^{DL} / 8) \right\rfloor$ for an extended cyclic prefix.

$N_{\text{RB}}^{\text{DL}}$ corresponds to the bandwidth of the radio channel, expressed as the number of Resource Blocks (RB).

The value of parameter N_g is supplied by the Master Information Block (MIB) message, which is included in the Physical Broadcast Channel (PBCH).

N_g can take the following values: $N_g \in \{1/6, 1/2, 1, 2\}$.

For the case of a structure with a type-2 time frame that corresponds to the Time Division Duplex (TDD) mode, the number of PHICH groups depends on the configuration that allocates each subframe to the downlink or uplink direction.

The number of PHICH groups is obtained with the formula $m_i \cdot N_{\text{PHICH}}^{\text{group}}$.

Table 7.3 provides the value of the multiplication factor m_i for each subframe, according to the frame configuration.

Frame configuration	Subframe									
	0	1	2	3	4	5	6	7	8	9
0	2	1	–	–	–	2	1	–	–	–
1	0	1	–	–	1	0	1	–	–	1
2	0	0	–	1	0	0	0	–	1	0
3	1	0	–	–	–	0	0	0	1	1
4	0	0	–	–	0	0	0	0	1	1
5	0	0	–	0	0	0	0	0	1	0
6	1	1	–	–	–	1	1	–	–	1

Table 7.3. *Value of the multiplication factor* m_i

7.3.3. *Error correction code*

The error correction code is a repetition code, the HI bit is repeated three times to constitute the bit sequence $b(0),...,b(M_{\text{bit}}-1)$, $M_{\text{bit}}=3$:

– 0, 0, 0 for the negative acknowledgement;

– 1, 1, 1 for the positive acknowledgement.

7.3.4. *Modulation*

The sequence of symbols $z(0),...,z(M_{\text{s}}-1)$, which corresponds to a Binary Phase-Shift Keying (BPSK) modulation, is generated from the sequence of bits $b(0),...,b(M_{\text{bit}}-1)$, where the number of symbols M_{s} equals 3.

7.3.5. *Scrambling and spreading*

The sequence of symbols $z(0),...,z(M_s-1)$ is repeated N_{SF}^{PHICH} times and multiplied by an orthogonal spreading sequence and by a Cell-Specific scrambling sequence to generate a block of symbols $d(0),d(1),...,d(M_{symb}-1)$, with:

$$M_{symb}=3\times N_{SF}^{PHICH}$$

– $N_{SF}^{PHICH}=4$, for the case of a normal cyclic prefix. An HI information generates 12 symbols;

– $N_{SF}^{PHICH}=2$, for the case of an extended cyclic prefix. An HI information generates 6 symbols.

The block of symbols $d(0),d(1),...,d(M_{symb}-1)$ is obtained with the following formula:

$$d(i)=w\left(i \bmod N_{SF}^{PHICH}\right)\cdot\left(1-2c(i)\right)\cdot z\left(\left\lfloor i/N_{SF}^{PHICH}\right\rfloor\right)$$

– $w\left(i \bmod N_{SF}^{PHICH}\right)$ represents the orthogonal sequence that allows to multiplex several PHICH physical channels in the same PHICH group (Table 7.4);

– $c(i)$ represents the Cell-Specific scrambling sequence, generated with a pseudo-random sequence that uses a 31-bit polynomial.

N_{PHICH}^{seq}	Normal cyclic prefix w(0), w(1), w(2), w(3)	Extended cyclic prefix w(0), w(1)
0	$[+1\ +1\ +1\ +1]$	$[+1\ +1]$
1	$[+1\ -1\ +1\ -1]$	$[+1\ -1]$
2	$[+1\ +1\ -1\ -1]$	$[+j\ +j]$
3	$[+1\ -1\ -1\ +1]$	$[+j\ -j]$
4	$[+j\ +j\ +j\ +j]$	–
5	$[+j\ -j\ +j\ -j]$	–
6	$[+j\ +j\ -j\ -j]$	–
7	$[+j\ -j\ -j\ +j]$	–

Table 7.4. *Values of the orthogonal sequence*

The initial value c_{init} of the scrambling sequence $c(i)$ is obtained with the following formula:

$$c_{\text{init}} = \left(\lfloor n_s/2 \rfloor + 1\right) \cdot \left(2N_{\text{ID}}^{\text{cell}} + 1\right) \cdot 2^9 + N_{\text{ID}}^{\text{cell}}$$

– n_s corresponds to the time slot number of the time frame;

– $N_{\text{ID}}^{\text{cell}}$ corresponds to the Physical-layer Cell Identity (PCI).

7.3.6. *Resource group alignment*

A Resource Element Group (REG) is made up of four resource elements, where each resource element contains a symbol.

The sequence of symbols $d(0),...,d(M_{\text{symb}} - 1)$ must be aligned with the size of the resource element group to produce the sequence of symbols $d^{(0)}(0),...,d^{(0)}(c \cdot M_{\text{symb}} - 1)$:

– $c = 1$ in case of a normal prefix;

– $c = 2$ in case on an extended prefix.

In the case of a normal cyclic prefix, each symbol of the sequence $z(0),...,z(M_s - 1)$ is represented by four scrambled symbols, and in this case, the alignment is not required:

$$d^{(0)}(i) = d(i), \text{ for } i = 0,..., M_{\text{symb}} - 1$$

In the case of an extended cyclic prefix, each symbol of the sequence, $z(0),...,z(M_s - 1)$ is represented by two scrambled signals.

To create a four-symbol block, zeros are added:

– before the block of scrambled symbols, if the parameter $N_{\text{PHICH}}^{\text{seq}}$ has an odd value;

– after the block if the parameter $N_{\text{PHICH}}^{\text{seq}}$ has an even value.

7.3.7. *Mapping on the spatial layers*

The set of symbols $d^{(0)}(0),...,d^{(0)}(c \cdot M_{\text{symb}}-1)$ is mapped on the set of symbols $x^{(0)}(i)$ ... $x^{(\upsilon-1)}(i)$, with $i = 0,1,...,M_{\text{symb}}^{\text{layer}}-1$, where υ corresponds to the number of spatial layers and $M_{\text{symb}}^{\text{layer}}$ corresponds to the number of symbols per spatial layer.

The number of spatial layers corresponds to the number of antenna ports.

When one antenna port is used (p0), the following relation is obtained:

$$x^{(0)}(i) = d^{(0)}(i)$$

When two antenna ports are used (p0 and p1), the following relations are obtained:

$$x^{(0)}(i) = d^{(0)}(2i)$$
$$x^{(1)}(i) = d^{(0)}(2i+1)$$

$$M_{\text{symb}}^{\text{layer}} = c.M_{\text{symb}}/2$$

When four antenna ports are used (p0 to p3), the following relations are obtained:

$$x^{(0)}(i) = d^{(0)}(4i)$$
$$x^{(1)}(i) = d^{(0)}(4i+1)$$
$$x^{(2)}(i) = d^{(0)}(4i+2)$$
$$x^{(3)}(i) = d^{(0)}(4i+3)$$

$$M_{\text{symb}}^{\text{layer}} = c.M_{\text{symb}}/4$$

7.3.8. *Precoding*

The precoding is performed in the set of symbols $x^{(0)}(i)$ ... $x^{(\upsilon-1)}(i)$ to generate the sequences $y^{(p)}(0),...,y^{(p)}(M_{\text{symb}}-1)$, where p corresponds to the antenna port number.

When one single antenna port is used (p0), no precoding is performed.

When two antenna ports are used (p0 and p1), the PHICH physical channel is transmitted with Space Frequency Block Coding (SFBC), a form of transmit diversity.

The symbols are transmitted using the following precoding:

– antenna port (0) $y^{(0)}(2i) = x^{(0)}(i)$ $y^{(0)}(2i+1) = x^{(1)}(i)$;

– antenna port (1) $y^{(1)}(2i) = -(x^{(1)}(i))*$ $y^{(1)}(2i+1) = (x^{(0)}(i))*$.

The asterisk symbol (*) represents the complex conjugate.

When four antenna ports are used (p0 to p3), the PHICH physical channel is transmitted with SFBC/FSTD (Frequency Shift Transmit Diversity).

The symbols are transmitted using the following precoding if:

– $(i + n_{\text{PHICH}}^{\text{group}}) \bmod 2 = 0$ for a normal cyclic prefix;

– $(i + \lfloor n_{\text{PHICH}}^{\text{group}}/2 \rfloor) \bmod 2 = 0$ for an extended cyclic prefix.

Antenna port (0) $y^{(0)}(4i) = x^{(0)}(i)$ $y^{(0)}(4i+1) = x^{(1)}(i)$

$y^{(0)}(4i+2) = x^{(2)}(i)$ $y^{(0)}(4i+3) = x^{(3)}(i)$

Antenna port (2) $y^{(2)}(4i) = -(x^{(1)}(i))*$ $y^{(2)}(4i+1) = (x^{(0)}(i))*$

$y^{(2)}(4i+2) = -(x^{(2)}(i))*$ $y^{(2)}(4i+3) = (x^{(2)}(i))*$

The symbols are transmitted using the following precoding if:

– $(i + n_{\text{PHICH}}^{\text{group}}) \bmod 2 = 1$ for a normal cyclic prefix;

– $(i + \lfloor n_{\text{PHICH}}^{\text{group}}/2 \rfloor) \bmod 2 = 1$ for an extended cyclic prefix.

Antenna Port (1) $y^{(1)}(4i) = x^{(0)}(i)$ $y^{(1)}(4i+1) = x^{(1)}(i)$

$$y^{(1)}(4i+2) = x^{(2)}(i) \qquad y^{(1)}(4i+3) = x^{(3)}(i)$$

Antenna port (3) $\quad y^{(3)}(4i) = -(x^{(1)}(i))* \qquad y^{(3)}(4i+1) = (x^{(0)}(i))*$

$$y^{(3)}(4i+2) = -(x^{(2)}(i))* \qquad y^{(3)}(4i+3) = (x^{(2)}(i))*$$

7.3.9. *Mapping on the resource elements*

The number of OFDM symbols used for the transport of the PHICH physical channel is configurable. It is indicated in the Master Information Block (MIB) message, included in the PBCH physical channel.

The duration of the PHICH physical channel is either normal or extended.

When the duration is normal, the PHICH physical channel is only present in the first OFDM symbol of each subframe.

When the duration is extended, the PHICH physical channel is present, subject to exceptions, in the first three OFDM symbols of each subframe.

The exception covers the following subframes, for which the PHICH physical channel is only present in the first two OFDM symbols:

– subframes 1 and 6 of the type-2 time frame, in TDD mode;

– MBMS Single Frequency Network (MBSFN) subframes.

The sequence $\bar{y}^{(p)}(0),...,\bar{y}^{(p)}(M_{\text{symb}}^{(0)}-1)$ of a PHICH group is the sum of all sequences of each PHICH physical channel: $\bar{y}^{(p)}(n) = \sum y_i^{(p)}(n)$.

For a normal cyclic prefix, the mapping of the PHICH group $\bar{y}_m^{(p)}(n)$ with the index m on the PHICH mapping unit $\tilde{y}_{m'}^{(p)}(n)$ with the index m' is defined by the relation $\tilde{y}_{m'}^{(p)}(n) = \bar{y}_m^{(p)}(n)$:

– $m' = m = 0,1,...,N_{\text{PHICH}}^{\text{group}} - 1$, for a type-1 time frame;

– $m' = m = 0,1,...,m_i.N_{\text{PHICH}}^{\text{group}} - 1$ for a type-2 time frame.

The values of the multiplication factor m_i are shown in Table 7.3.

For an extended cyclic prefix, the mapping of the PHICH group $\bar{y}_m^{(p)}(n)$ with the index m and $m+1$ on the PHICH mapping unit $\tilde{y}_{m'}^{(p)}(n)$ with the index m' is defined by the relation $\tilde{y}_{m'}^{(p)}(n) = \bar{y}_m^{(p)}(n) + \bar{y}_{m+1}^{(p)}(n)$.

– $m' = m/2$;

– $m = 0,2,...,N_{\text{PHICH}}^{\text{group}} - 2$, for a type-1 time frame;

– $m = 0,2,...,m_i.N_{\text{PHICH}}^{\text{group}} - 2$, for a type-2 time frame.

Let $z^{(p)}(i) = \left\langle \tilde{y}^{(p)}(4i), \tilde{y}^{(p)}(4i+1), \tilde{y}^{(p)}(4i+2), \tilde{y}^{(p)}(4i+3) \right\rangle$ constituting a quadruplet of index i, with $i = 0,1,2,$ for the antenna port p, the mapping of quadruplet $z^{(p)}(i)$ is performed on the resource elements represented by pair $(k',l')_i$.

The index l'_i corresponds to the number of the OFDM symbol, in the time domain:

– $l'_i = 0$, for the normal duration of the PHICH physical channel;

– $l'_i = \left(\lfloor m'/2 \rfloor + i + 1\right) \bmod 2$, for subframes 1 and 6 of type-2 time frame, in TDD mode, and MBSFN subframes, when using an extended duration of the PHICH physical channel;

– $l'_i = i$, for the other subframes, when using an extended duration of the PHICH physical channel.

The index k'_i corresponds to the number of the resource element that represents the resource element group in the frequency domain.

The index k'_i is represented by number $\bar{n}_i$ from the resource element group.

For subframes 1 and 6 of type-2 time frame, in TDD mode, and MBSFN subframes, when using an extended duration of the PHICH physical channel:

$$\bar{n}_i = \begin{cases} \left(\left\lfloor N_{\mathrm{ID}}^{\mathrm{cell}} \cdot n_{l_i'} / n_1 \right\rfloor + m'\right) \bmod n_{l_i'} & i = 0 \\ \left(\left\lfloor N_{\mathrm{ID}}^{\mathrm{cell}} \cdot n_{l_i'} / n_1 \right\rfloor + m' + \left\lfloor n_{l_i'} / 3 \right\rfloor\right) \bmod n_{l_i'} & i = 1 \\ \left(\left\lfloor N_{\mathrm{ID}}^{\mathrm{cell}} \cdot n_{l_i'} / n_1 \right\rfloor + m' + \left\lfloor 2 n_{l_i'} / 3 \right\rfloor\right) \bmod n_{l_i'} & i = 2 \end{cases}$$

For the other subframes:

$$\bar{n}_i = \begin{cases} \left(\left\lfloor N_{\mathrm{ID}}^{\mathrm{cell}} \cdot n_{l_i'} / n_0 \right\rfloor + m'\right) \bmod n_{l_i'} & i = 0 \\ \left(\left\lfloor N_{\mathrm{ID}}^{\mathrm{cell}} \cdot n_{l_i'} / n_0 \right\rfloor + m' + \left\lfloor n_{l_i'} / 3 \right\rfloor\right) \bmod n_{l_i'} & i = 1 \\ \left(\left\lfloor N_{\mathrm{ID}}^{\mathrm{cell}} \cdot n_{l_i'} / n_0 \right\rfloor + m' + \left\lfloor 2 n_{l_i'} / 3 \right\rfloor\right) \bmod n_{l_i'} & i = 2 \end{cases}$$

$n_{l'}$ represents the number of resource element groups not allocated to the Physical Control Format Indicator Channel (PCFICH).

The numbering of resource element groups ranges from 0 to $n_{l'} - 1$.

Figure 7.6 shows an example of group positioning for the PHCIH physical channel, obtained from the following assumptions:

– FDD mode;

– normal cyclic prefix;

– normal duration of physical channel PHICH: $l_i' = 0$.

$N_{\mathrm{ID}}^{\mathrm{cell}} = 0$

The number of resource blocks in the frequency domain $N_{\mathrm{RB}}^{\mathrm{DL}}$ is equal to 6, which corresponds to a bandwidth in the radio channel with 1.4-MHz.

The number of resource element groups is equal to $n_{l_i'} = n_0 = 8$.

The number of PHICH groups $N_{\mathrm{PHICH}}^{\mathrm{group}}$ is equal to 1 ($N_g = 1$).

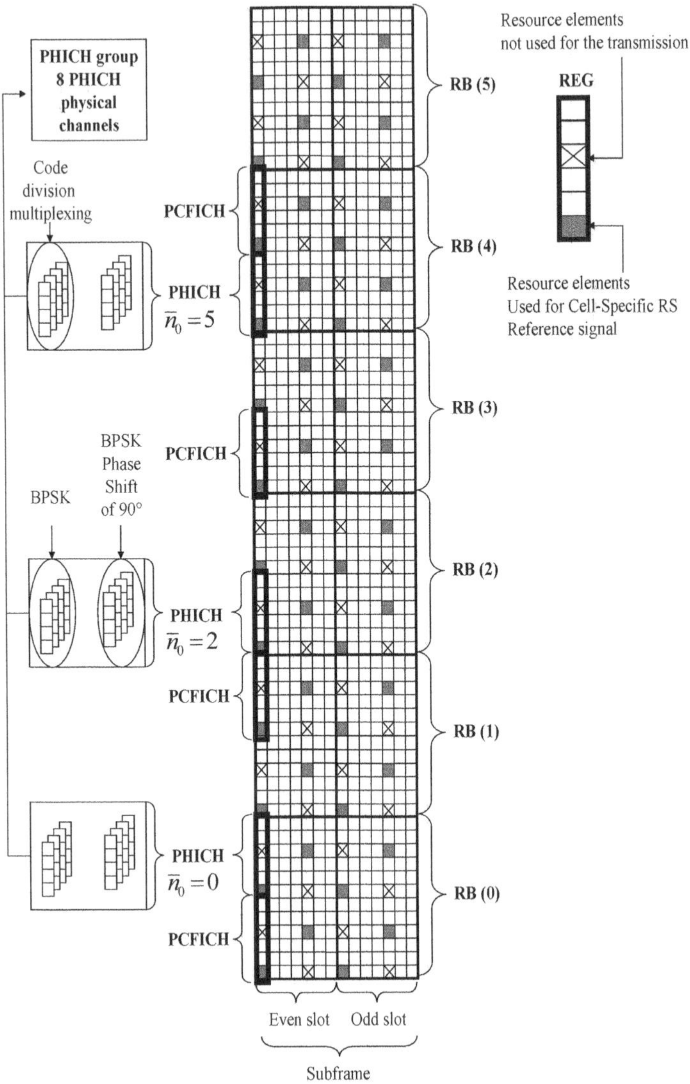

Figure 7.6. *Mapping of the PHICH physical channels*

These are the resource element groups used:

– $\bar{n}_0 = 0$ is represented by the index pair $(k' = 1, l' = 0)$;

– $\overline{n}_1 = 2$ is represented by the index pair $(k' = 4, l' = 0)$;

– $\overline{n}_2 = 5$ is represented by the index pair $(k' = 8, l' = 0)$.

7.3.10. *Assignment of the PHICH physical channel*

In the FDD mode, the mobile analyzes the PHICH physical channel with a 4-ms time delay k_{PHICH}, as compared with the transmission of data in the PUSCH physical channel.

In the TDD mode, the time delay value k_{PHICH} is a function of the configuration for the time frame and for the subframe number (Table 7.5).

Type of configuration	Number of the subframe									
	0	1	2	3	4	5	6	7	8	9
0			4	7	6			4	7	6
1			4	6				4	6	
2			6					6		
3			6	6	6					
4			6	6						
5			6							
6			4	6	6			4	7	

Table 7.5. *Value of the time delay* k_{PHICH}

The PHICH physical channel allocated to the mobile is identified by the index pair $(n_{PHICH}^{group}, n_{PHICH}^{seq})$, related to the group number and the sequence number, and calculated with the following formulas:

$$n_{PHICH}^{group} = (I_{PRB\_RA} + n_{DMRS}) \bmod N_{PHICH}^{group} + I_{PHICH} N_{PHICH}^{group}$$

$$n_{PHICH}^{seq} = (\lfloor I_{PRB\_RA} / N_{PHICH}^{group} \rfloor + n_{DMRS}) \bmod 2N_{SF}^{PHICH}$$

I_{PRB_RA} is the lowest index of the resource blocks allocated in the PUSCH physical channel.

In Release 10, the value of I_{PRB_RA} is increased by one unit for the second transport block:

– n_{DMRS} is the value of cyclic shift for Demodulation Reference Signal (DM-RS) , reported in the Downlink Control Information (DCI), formats 0 and 4;

– N_{SF}^{PHICH} is the value of the spreading factor for the orthogonal sequence (4 for FDD mode, 2 for TDD mode);

– I_{PHICH} has a value equal to 1, for a transmission of the PUSCH physical channel in subframes 4 and 9, in TDD mode, or a value equal to 0 for the other transmission instances.

7.4. PDCCH physical channel

The Physical Downlink Control Channel (PDCCH) transmits the Downlink Control Information (DCI) for one or several User Equipment (UE):

– resource allocation, modulation and coding schemes, for the data included in the Physical Downlink Shared Channel (PDSCH) and in the Physical Uplink Shared Channel (PUSCH);

– transmit power control of the Physical Uplink Control Channel (PUCCH) and the PUSCH physical channel.

For each subframe, the PDCCH physical channel takes the first, the first two, or the first three Orthogonal Frequency-Division Multiplexing (OFDM) symbols.

The number of OFDM symbols allocated to the PDCCH physical channel is indicated in the Physical Control Format Indicator Channel (PCFICH).

The processing associated with the PDCCH physical channel is summarized in Figure 7.7.

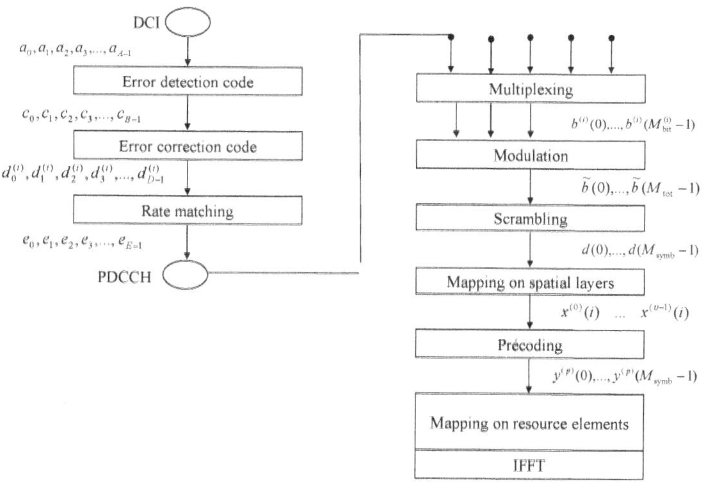

Figure 7.7. *Processing associated with the PDCCH physical channel*

7.4.1. *DCI information*

The format of the DCI information is a function of the Radio Network Temporary Identifier (RNTI) and the transmission mode (Tables 7.6–7.8).

The Cell RNTI (C-RNTI) and the Semi-Persistent Scheduling C-RNTI (SPS C-RNTI) allow us to retrieve the information regarding the description of the resource allocated to the mobile for the uplink and downlink direction.

The System Information RNTI (SI-RNTI) allows us to retrieve the System Information Block (SIB).

The Paging RNTI (P-RNTI) allows us to retrieve the data related with paging information.

The Random Access RNTI (RA-RNTI) allows us to retrieve the information transmitted during the random access procedure.

The Temporary Cell RNTI (TC-RNTI) allows us to retrieve the information exchanged during the connection procedure following the random access procedure.

The Transmit Power Control RNTI (TPC-RNTI) allows us to retrieve the information concerning the power control of the PUSCH and PUCCH physical channels.

Types of RNTI	Types of format DCI
SI-RNTI, P-RNTI, RA-RNTI	1A, 1C
C-RNTI, SPS C-RNTI	0, 1, 1A, 1B, 1D, 2, 2A, 2B, 2C, 4
TPC-RNTI	3, 3A

Table 7.6. *Correspondence between RNTI types and DCI formats*

Modes of transmission	Types of format DCI
Mode 1	1, 1A
Mode 2	1, 1A
Mode 3	1A, 2A
Mode 4	1A, 2
Mode 5	1A, 1D
Mode 6	1A, 1B
Mode 7	1, 1A
Mode 8	1A, 2B
Mode 9	1A, 2C

Table 7.7. *Correspondence between transmission modes and DCI formats: downlink direction*

Modes of transmission	Types of format DCI
Mode 1	0
Mode 2	4

Table 7.8. *Correspondence between transmission modes and DCI formats: uplink direction*

Formats 0 and 4 allocate the resource in the PUSCH physical channel and adjust the power of the PUSCH physical channel:

– format 0 allocates the resource for one transport block;

– format 4 allocates the resource for two transport blocks, when using the Multiple Input Multiple Output (MIMO) mechanism from Release 10.

Formats 1, 1A, 1B, 1D define the resources in the PDSCH physical channel, for one transport block, and adjust the power of the PUCCH physical channel:

– format 1A introduces resource allocation in the PDSCH physical channel , for the information concerning the random access, the paging and the SIB information;

– format 1B introduces information concerning the Precoding Matrix Indicator (PMI), proposed for the PDSCH physical channel;

– format 1D introduces information concerning the multi-user MIMO mechanism, for the transmission mode 5.

Format 1C allocates resources in the PDSCH physical channel solely for information concerning the random access, the paging and the SIB information.

Formats 2, 2A, 2B, 2C allocate the resources in the PDSCH physical channel, for two transport blocks, and adjust the power of the PUCCH physical channel:

– format 2A deletes the information concerning PMI precoding;

– format 2B is related to transmission mode 8, introduced in Release 9;

– format 2C is related to transmission mode 9, introduced in Release 10.

Formats 3 and 3A are solely used for the power control of PUCCH and PUSCH physical channels:

– a two-bit power adjustment is used for format 3;

– a one-bit power adjustment is used for format 3A.

Table 7.9 summarizes DCI information of format type 1 and 2 in regard to scheduling of the downlink direction.

Information	DCI								
	1	1A	1B	1C	1D	2	2A	2B	2C
Carrier indicator	■	■	■		■	■	■	■	■
Gap value				■					
Localized/Distributed VRB assignment flag		■	■		■				
Flag for format0/format1A differentiation		■							
Resource allocation	■					■	■	■	■
Resource block assignment	■	■	■	■	■	■	■	■	■
Modulation and coding scheme	■	■	■	■	■	■	■	■	■
HARQ process	■	■	■		■	■	■	■	■
New data indicator	■	■	■		■	■	■	■	■
Redundancy version	■	■	■		■	■	■	■	■
TPC command for PUCCH	■	■	■		■	■	■	■	■
Downlink Assignment Index	■	■	■		■	■	■	■	■
SRS request		■						■	■
TPMI information for precoding			■		■				
PMI confirmation for precoding			■						
MCCH change notification				■					
Downlink power offset					■				
Transport block to codeword swap flag						■	■		
Precoding information						■	■		
Scrambling identity								■	
Antenna port(s), scrambling identity and number of layers									■

Table 7.9. *DCI information of format type 1 and 2*

Carrier indicator: this field indicates the number of the radio channel in the aggregation. A value equal to 0 is allocated to the primary channel Pcell.

Gap value: this field indicates the parameter $N_{\text{gap},1}$ or $N_{\text{gap},2}$, which is used for the mapping of the Virtual Resource Blocks (VRB).

Localized/Distributed VRB assignment flag: this field indicates if the mapping of the virtual resource blocks is either localized or distributed.

Flag for format0/format1A differentiation: this field allows us to determine the format 0 or 1A of the DCI.

Resource allocation: this field indicates the type of resource allocation; type 0 or type 1, which defines the bit mapping of the Resource Block Groups (RBGs) allocated to the mobile.

Resource block assignment: this field shows the location in the frequency domain, of the Resource Blocks (RB) allocated.

Modulation and coding scheme (MCS): this field supplies an index that corresponds to a modulation and coding scheme.

HARQ process: this field indicates the number of the Hybrid Automatic Repeat reQuest (HARQ) process.

New data indicator (NDI): this field indicates whether the data is new or is part of a retransmission.

Redundancy version (RV): this field indicates the number of the redundancy version.

TPC command for PUCCH: this field supplies the incremental value for power control of the PUCCH physical channel.

Downlink Assignment Index: this field, used in the Time Division Duplex (TDD) mode, indicates the number of transmissions for the downlink direction whose evolved Node B (eNB) entity waits for the acknowledgement.

SRS request: this field indicates that the mobile must generate the Sounding Reference Signal (SRS).

TPMI information for precoding: this field indicates the index for the Transmitted Precoding Matrix Indicator (TPMI).

PMI confirmation for precoding: this field indicates if precoding uses the TPMI or PMI information previously transmitted by the mobile.

MCCH change notification: this field indicates a modification of the Multicast Control Channel (MCCH).

Downlink power offset: this field indicates, in the case of the multi-user MIMO transmission mode, the power offset (0 dB or – 3 dB) as compared with that transmitted with a single user.

Transport block to codeword swap flag: this field defines the relationship between the code word and the transport block.

Precoding information: this field supplies a table index corresponding to a precoding value.

Scrambling identity: this field indicates the scrambling identity and makes it possible, in the case of the multi-user MIMO transmission mode, for two mobiles to share the same antenna port.

Antenna port(s), scrambling identity and number of layers: this field indicates the value of a table pointer defining the number of spatial layers, the number of antenna ports and the scrambling identity.

Table 7.10 summarizes the DCI information for format 0 and 4 in regard to the scheduling of the uplink direction.

Information	DCI	
	0	4
Carrier indicator	■	■
Flag for format0/format1A differentiation	■	
Frequency hopping flag	■	
Resource allocation	■	■
Resource block assignment	■	■
Modulation and coding scheme	■	■
New data indicator	■	■
Redundancy version	■	■
TPC command for PUSCH	■	■
Cyclic shift for DM RS and OCC index	■	■
UL index	■	■
Downlink Assignment Index	■	■
CSI request	■	■
SRS request		■
Precoding information		■

Table 7.10. *DCI information for format 0 and 4*

Frequency hopping flag: this field indicates if a frequency hopping procedure must be applied to the PUSCH physical channel.

TPC command for PUSCH: this field supplies the incremental value for the power control of the PUSCH physical channel.

Cyclic shift for DM RS and OCC index: this field supplies a table pointer defining the value of the cyclic shift used for the Demodulation Reference Signal (DM-RS). Release 10 included in the Orthogonal Cover Code (OCC) table, which allows the spatial multiplexing of several mobiles for the multi-user MIMO transmission mode.

UL index: this field is solely included for the TDD mode and for the configuration 0. It indicates the instant in which the following operations must be applied:

– the power command of the PUSCH physical channel;

– the Channel State Information (CSI) concerning the reports of the signal received over the PDSCH physical channel;

– the transmission over the PUSCH physical channel.

CSI request: this field indicates that CSI aperiodic reports must be recovered for a radio channel or for a set of radio channels that are part of an aggregation.

7.4.2. *Error detection code*

Error detection is obtained with a Cyclic Redundancy Check (CRC) code.

The CRC structure is the remainder when the transport block $a_0, a_1, a_2, a_3, ..., a_{A-1}$, that corresponds to the DCI information , is divided by a generator polynomial, whose remainder $p_0, p_1, p_2, p_3, ..., p_{L-1}$ (L=16) constitutes the cyclic redundancy bits.

The concatenation of the transport block $a_0, a_1, a_2, a_3, ..., a_{A-1}$ and the 16 bits from the cyclic redundancy $p_0, p_1, p_2, p_3, ..., p_{L-1}$ constitute the structure $b_0, b_1, b_2, b_3, ..., b_{B-1}$.

Since the PDCCH physical channel carries the DCI information specifically to a mobile or shared by several mobiles, an EXCLUSIVE OR is performed between the CRC structure and the RNTI $x_{rnti,0}, x_{rnti,1}, \ldots, x_{rnti,15}$ to obtain the sequence $c_0, c_1, c_2, c_3, \ldots, c_{B-1}$:

– $c_k = b_k$, for $k = 0, 1, 2, \ldots, A\text{-}1$;

– $c_k = \left(b_k + x_{rnti,k-A}\right) \bmod 2$, for $k = A, A+1, A+2, \ldots, A+15$.

In a case where it is necessary to configure the number of the antenna for the mobile transmission, for the format 0, an antenna selection mask is applied $x_{AS,0}, x_{AS,1}, \ldots, x_{AS,15}$ (Table 7.11).

Number of antenna	Antenna selection mask $x_{AS,0}, x_{AS,1}, \ldots, x_{AS,15}$
0	0,0,0,0,0,0,0,0,0,0,0,0,0,0,0,0
1	0,0,0,0,0,0,0,0,0,0,0,0,0,0,0,1

Table 7.11. *Antenna selection mask*

7.4.3. *Error correction code*

The error correction code is a convolutional code that produces three sequences $d_0^{(i)}, d_1^{(i)}, d_2^{(i)}, d_3^{(i)}, \ldots, d_{D-1}^{(i)}$ with $i = 0, 1$, and 2 and D corresponding to the number of bits of the sequence.

7.4.4. *Rate matching*

The propagation condition determines the coding scheme retained. Rate matching determines the sequence with the desired coding rate.

The three sequences emitted from the convolutional code are interleaved and then stored in a circular memory. Bits of the circular memory are selected or punctured to constitute the output sequence $e_0, e_1, e_2, e_3, \ldots, e_{E-1}$, where E corresponds to the number of bits of the sequence.

7.4.5. *Multiplexing*

The DCI information is transmitted in an aggregation of the Control Channel Elements (CCE).

Each control channel element occupies nine Resource Element Groups (REG).

Table 7.12 summarizes the structure for the diverse format types of the PDCCH physical channel.

Types of PDCCH format	Number of CCE	Number of REG	Number of RE	Number of bits
0	1	9	36	72
1	2	18	72	144
2	4	36	144	288
3	8	72	288	576

Table 7.12. *Types of the format of the PDCCH physical channel*

7.4.6. *Scrambling*

The bit sequence $b^{(i)}(0),...,b^{(i)}(M_{\text{bit}}^{(i)}-1)$, $M_{\text{bit}}^{(i)}$ which corresponds to the number of bits in channel i, is scrambled with the sequence $c(i)$.

$i=0,...n_{\text{PDCCH}}$, n_{PDCCH} corresponding to the number of the PDCCH physical channel.

This results in the bit sequence $\tilde{b}(0),...,\tilde{b}(M_{\text{tot}}-1)$, M_{tot} corresponding to the number of bits of the sequence, $M_{\text{tot}}=\sum M_{\text{bit}}^{(i)}$.

$$\tilde{b}(i)=\left(b(i)+c(i)\right)\text{mod}\,2$$

The scrambling sequence is built from a pseudo-random sequence that uses a 31-bit polynomial.

The initial value c_{init} of the scrambling sequence is calculated with the following formula:

$$c_{\text{init}}=\lfloor n_s/2\rfloor 2^9+N_{\text{ID}}^{\text{cell}}$$

– n_s is the number of the time slot in the time frame;

– N_{ID}^{cell} corresponds to the number of the Physical-layer Cell Identity (PCI).

7.4.7. *Modulation*

The sequence of symbols $d(0),...,d(M_{symb}-1)$, which corresponds to a Quadrature Phase-Shift Keying (QPSK) modulation, is obtained from the sequence of bits $\tilde{b}(0),...,\tilde{b}(M_{tot}-1)$, where a symbol corresponds to two bits.

7.4.8. *Mapping on the spatial layers*

The set of bits symbols $d(0),...,d(M_{symb}-1)$ is mapped to the set $x^{(0)}(i)$ ... $x^{(\upsilon-1)}(i)$, with $i=0,1,...,M_{symb}^{layer}-1$, where υ corresponds to the number of spatial layers and M_{symb}^{layer} .corresponds to the number of symbols per layer.

The number of spatial layers corresponds to the number of antenna ports.

When one antenna port (p0) is used, the following relation is obtained:

$$x^{(0)}(i)=d^{(0)}(i).$$

When two antenna ports (p0 and p1) are used, the following relations are obtained:

$$x^{(0)}(i)=d^{(0)}(2i)$$
$$x^{(1)}(i)=d^{(0)}(2i+1)$$

$$M_{symb}^{layer}=M_{symb}/2$$

When four antenna ports (p0 to p3) are used, the following relations are obtained:

$$x^{(0)}(i)=d^{(0)}(4i)$$

$$x^{(1)}(i)=d^{(0)}(4i+1)$$

$$x^{(2)}(i) = d^{(0)}(4i+2)$$

$$x^{(3)}(i) = d^{(0)}(4i+3)$$

$$M_{\text{symb}}^{\text{layer}} = M_{\text{symb}}/4$$

7.4.9. *Precoding*

The precoding is applied over the set of symbols $x^{(0)}(i)$ ... $x^{(\upsilon-1)}(i)$ to generate the sequences $y^{(p)}(0),...,y^{(p)}(M_{\text{symb}}-1)$, where p corresponds to the number of antenna port.

When one single antenna port is used (p0), no precoding is performed.

When two antenna ports are used (p0 and p1), the PDCCH physical channel is transmitted with Space Frequency Block Coding (SFBC), a form of transmit diversity.

The symbols are transmitted using the following precoding:

– antenna Port (0) $y^{(0)}(2i) = x^{(0)}(i)$ $\quad y^{(0)}(2i+1) = x^{(1)}(i)$;

– antenna Port (1) $y^{(1)}(2i) = -(x^{(1)}(i))*$ $\quad y^{(1)}(2i+1) = (x^{(0)}(i))*$.

When four antenna ports are used (p0 to p3), the PDCCH physical channel is transmitted with SFBC/FSTD (Frequency Shift Transmit Diversity).

The symbols are transmitted using the following precoding:

– antenna Port (0) $y^{(0)}(4i) = x^{(0)}(i)$ $\quad y^{(0)}(4i+1) = x^{(1)}(i)$;

– antenna Port (1) $y^{(1)}(4i+2) = x^{(2)}(i)$ $\quad y^{(1)}(4i+3) = x^{(3)}(i)$;

– antenna port (2) $y^{(2)}(4i) = -(x^{(1)}(i))*$ $\quad y^{(2)}(4i+1) = (x^{(0)}(i))*$;

– antenna port (3) $y^{(3)}(4i+2) = -(x^{(3)}(i))*$ $\quad y^{(3)}(4i+3) = (x^{(2)}(i))*$.

7.4.10. *Mapping on the resource elements*

The mapping on the resource elements is performed with a quadruplet of symbols $z^{(p)}(i) = \langle y^{(p)}(4i), y^{(p)}(4i+1), y^{(p)}(4i+2), y^{(p)}(4i+3) \rangle$, where i corresponds to the quadruplet index and p to the antenna port.

The quadruplet blocks $z^{(p)}(0),...,z^{(p)}(M_{\text{quad}}-1)$, $M_{\text{quad}} = M_{\text{symb}}/4$, are subject to a permutation, then to a cyclic shift to obtain the sequence $\overline{w}^{(p)}(0),...,\overline{w}^{(p)}(M_{\text{quad}}-1)$.

The mapping of the quadruplets $\overline{w}^{(p)}(i)$ is performed on the resource element groups not allocated to the Physical Control Format Indicator Channel (PCFICH) and to the Physical HARQ Indicator Channel (PHICH).

The resource element group is represented by the index pair (k', l').

The index k' corresponds to the resource element number that represents the resource element group in the frequency domain.

The index l' corresponds to the number of the OFDM symbol in the time domain.

The mapping of quadruplets starts with the index pair $(k' = 0, l' = 0)$.

The mapping is performed by increasing index l' until it is equal to the number of OFDM symbols allocated to the PDCCH physical channel.

The mapping continues by increasing index k' by one unit and shifting index l' as described before.

The mapping stops when $k' = N_{\text{RB}}^{\text{DL}} \cdot N_{\text{sc}}^{\text{RB}}$:

– $N_{\text{RB}}^{\text{DL}}$ corresponds to the bandwidth of radio channel, expressed as the number of resource blocks;

– $N_{\text{sc}}^{\text{RB}}$ corresponds to the size of the resource block in the frequency domain, expressed as the number of subcarriers.

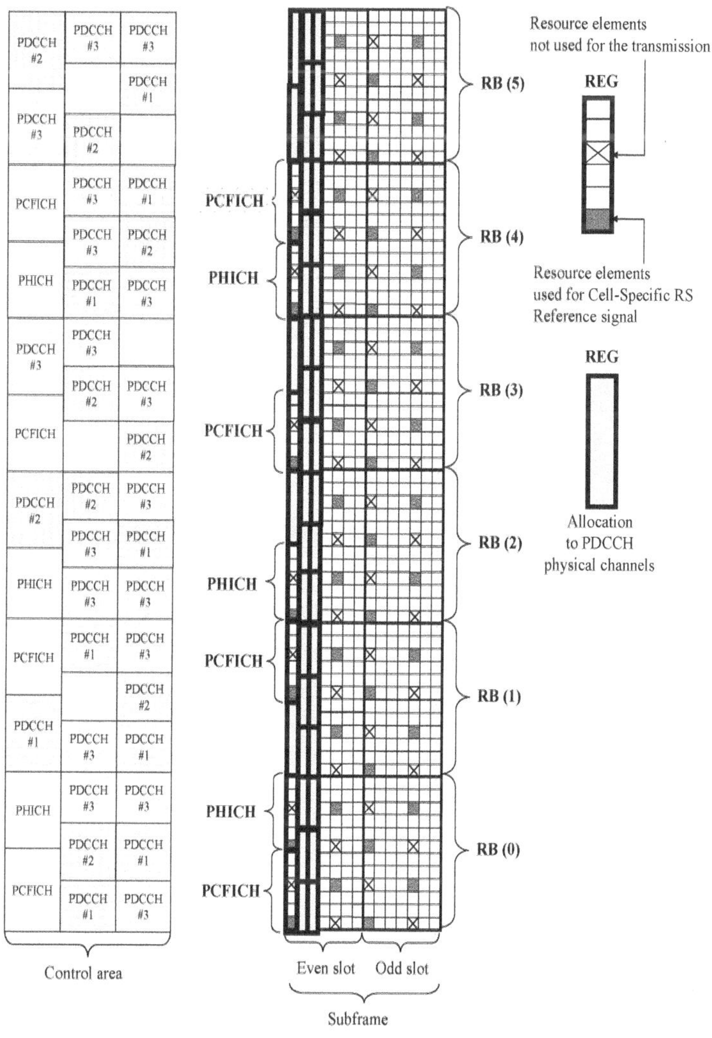

Figure 7.8. *Mapping of the PDCCH physical channels*

Figure 7.8 shows the position of quadruplets corresponding to several PDCCH physical channels, to which the first three OFDM symbols of the subframe are allocated:

– PDCCH physical channel #1 consists of 9 resource element groups;

– PDCCH physical channel #2 consists of 9 resource element groups;

– PDCCH physical channel #3 consists of 18 resource element groups.

For Release 10, as part of the aggregation of the Component Carrier (CC), the physical channel PDCCH can either carry the scheduling information for the radio channel where it is transmitted or allocate resources for other radio channels in case of an inter-carrier scheduling.

The inter-carrier scheduling allows deleting inter-cell interference over the PDCCH physical channel by allocating to the adjacent cells, resources over different radio channels.

The inter-carrier scheduling can only be applied to the secondary channels Scell, since the primary channel PCell systematically has a PDCCH physical channel allocated.

The inter-carrier scheduling can be established in accordance with two scenarios (Figure 7.9):

– the PDCCH physical channel is carried only by the primary radio channel Pcell;

– the PDCCH physical channel is carried by the primary radio channel Pcell and by a secondary radio channel SCell.

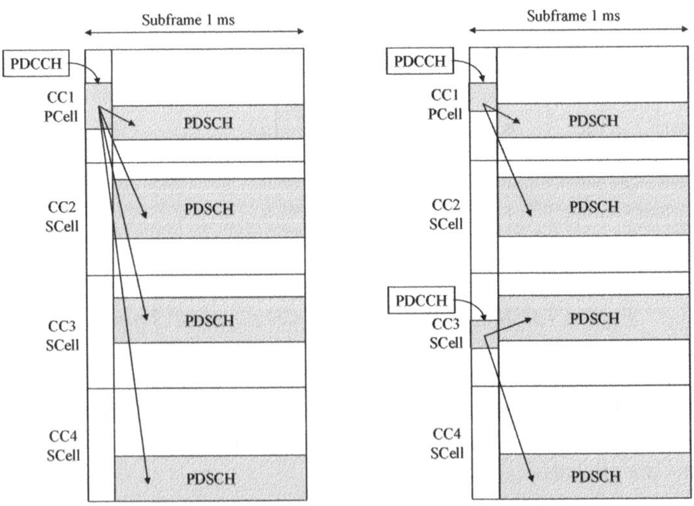

Figure 7.9. *Inter-carrier scheduling*

7.4.11. *Allocation of physical channel PDCCH*

The mobile must carry out a blind decoding for a set of candidate PDCCH physical channels, the candidate is defined by a search space.

The search spaces have been created to limit the number of PDCCH physical channels that the mobile must analyze.

There is a search space that is common to all mobiles and a specific search space for each mobile.

The common search space is used to signal the information concerning the random access, the paging, the SIB information and the power control.

The common search space can also be used to signal dedicated information to the mobile, if the specific search space is saturated.

The common search space contains the DCI information according to formats 0, 1A, 1C, 3 and 3A.

The specific search space is used to signal dedicated information to a mobile.

The common search space contains the DCI information according to formats 0, 1, 1A, 1B, 1D, 2, 2A, 2B, 2C, 4.

Both types of search space are defined by the size and by the aggregation levels of the control channel elements (Table 7.13).

The common search space starts up in the first control channel element.

The tracking of the specific search space by a mobile depends on the RNTI value of the mobile and on the time slot number inside the time frame.

Type of search space	Aggregation level	Size	Number of candidates
Common	1	6	6
	2	12	6
	4	8	2
	8	16	2
Type of search space	Aggregation level	Size	Number of candidates
Specific	4	16	4
	8	16	2

Table 7.13. *Properties of the search spaces*

Tables 7.14–7.17 provide, for the downlink direction and according to the type of RNTI, the relations between the following parameters:

– the DCI format;

– the type of search space;

– the transmission mode.

DCI	Search space	Transmission scheme
1C	Common	Antenna port 0 or transmit diversity
1A	Common	Antenna port 0 or transmit diversity

Table 7.14. *Decoding of the PDCCH physical channel the PDSCH physical channel is configured by the SI-RNTI, P-RNTI, RA-RNTI identifiers*

DCI	Search space	Transmission scheme
1A	Common/Specific	Antenna port 0 or transmit diversity
1	Common/Specific	Antenna port 0 or transmit diversity

Table 7.15. *Decoding of the PDCCH physical channel the PDSCH physical channel is configured by the TC-RNTI identifier*

Mode	DCI	Search space	Transmission scheme
1	1A	Common/Specific	Antenna port 0
	1	Specific	
2	1A	Common/Specific	Transmit diversity
	1	Specific	
3	1A	Common/Specific	Transmit diversity
	2A	Specific	Open loop MIMO
4	1A	Common/Specific	Transmit diversity
	2	Specific	Closed loop MIMO
5	1A	Common/Specific	Transmit diversity
	1D	Specific (1)	Multi-user MIMO
6	1A	Common/Specific	Transmit diversity
	1B	Specific (1)	Closed loop MIMO one layer
7	1A	Common/Specific	Transmit diversity or antenna port 0
	1	Specific	Beamforming, antenna port 5
8	1A	Common/Specific	Transmit diversity/antenna port 0
	2B	Specific	Antenna Ports 7 and 8
9	1A	Common/Specific	Transmit diversity/antenna port 0
	2C	Specific	MIMO, antenna ports 7 to 14 (3)

Table 7.16. *Decoding of the PDCCH physical channel the PDSCH physical channel is configured by the C-RNTI identifier*

Mode	DCI	Search space	Transmission scheme
1	1A	Common/Specific	Antenna port 0
	1	Specific	
2	1A	Common/Specific	Transmit diversity
	1	Specific	
3	1A	Common/Specific	Transmit diversity
	2A	Specific	
4	1A	Common/Specific	Transmit diversity
	2	Specific	
5	1A	Common/Specific	Transmit diversity
6	1A	Common/Specific	Transmit diversity
7	1A	Common/Specific	Antenna port 5
	1	Specific	
8	1A	Common/Specific	Antenna port 7
	2B	Specific	Antenna port 7 or 8
9	1A	Common/Specific	Antenna port 7
	2C	Specific	Antenna port 7 or 8

Table 7.17. *Decoding of the PDCCH physical channel the PDSCH physical channel is configured by the SPS C-RNTI identifier*

For the case of inter-carrier scheduling, the eNB entity defines the use of the PDCCH physical channel in the RRC message *ConnectionReconfiguration*, transmitted in the primary channel PCell and possibly in the secondary channel SCell.

7.5. PDSCH physical channel

The Physical Downlink Shared Channel (PDSCH) transmits the Downlink Shared Channel (DL-SCH) and the Paging Channel (PCH).

The DL-SCH transport channel contains the traffic data (unicast IP packets) and the control data (RRC messages) that correspond to the following messages:

– the System Information Block (SIB);

– the messages of common or dedicated control;

– the transport messages for Non-Access Stratum (NAS) signaling.

The DL-SCH transport channel can appear as one or two transport blocks.

The PCH transport channel contains the RRC control data that corresponds to the paging message.

The processing associated with the PDSCH physical channel is summarized in Figure 7.10.

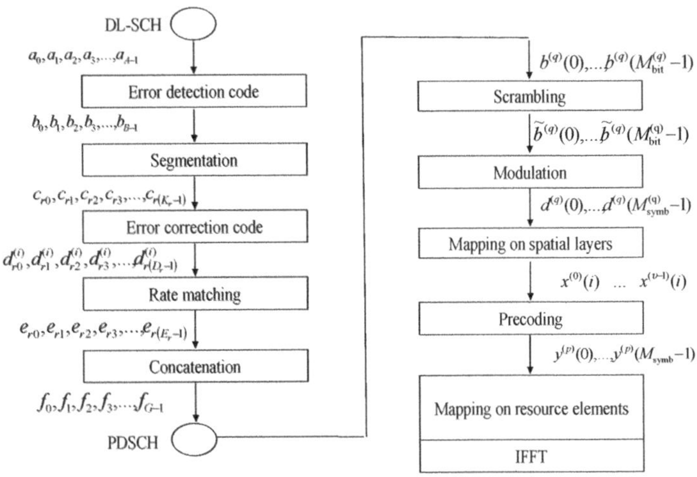

Figure 7.10. *Processing associated with the PDSCH physical channel*

7.5.1. *Error detection code*

The error detection is obtained with a Cyclic Redundancy Check (CRC) code.

The CRC structure is the remainder when the transport block $a_0, a_1, a_2, a_3, ..., a_{A-1}$ is divided by a generator polynomial, whose remainder $p_0, p_1, p_2, p_3, ..., p_{L-1}$ constitutes the cyclic redundancy bits.

The concatenation of the transport block $a_0, a_1, a_2, a_3, ..., a_{A-1}$ and of the 24 bits of the cyclic redundancy $p_0, p_1, p_2, p_3, ..., p_{L-1}$ constitute the input structure $b_0, b_1, b_2, b_3, ..., b_{B-1}$ of the segmentation.

7.5.2. *Segmentation*

The maximum size of the block processed by the error correction code is equal to 6144 bits.

If the block $b_0, b_1, b_2, b_3, ..., b_{B-1}$ is of a size larger than this value, it must be segmented and each segment must have its own CRC code, whose length equals 24 bits to constitute the sequence $c_{r0}, c_{r1}, c_{r2}, c_{r3}, ..., c_{r(K_r-1)}$, where r and K_r correspond respectively to the segment number and to the number of the segment bits

7.5.3. *Error correction code*

When the segmentation is performed over the sequence $b_0, b_1, b_2, b_3, ..., b_{B-1}$, the error correction mechanism is applied over each segment.

A turbo code handles error correction; it consists of a Parallel Concatenated Convolutional Code (PDCCC) with double convolution and a Quadratic Permutation Polynomial (QPP) as the interleaving (Figure 7.11).

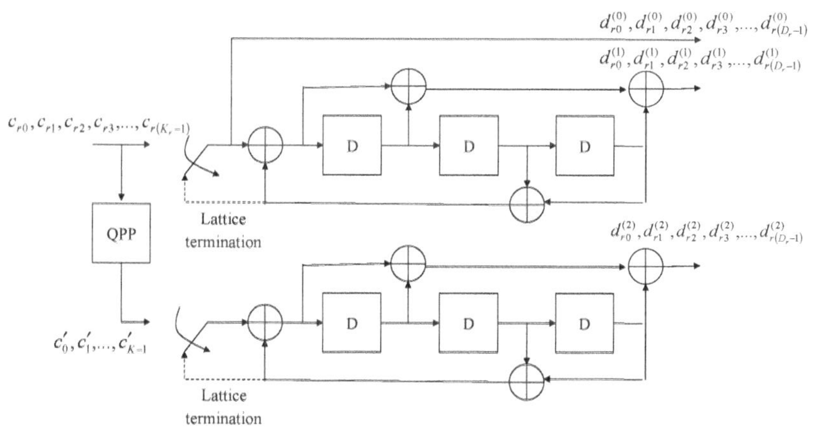

Figure 7.11. *Turbo code*

The error correction code produces three sequences $d_{r0}^{(i)}, d_{r1}^{(i)}, d_{r2}^{(i)}, d_{r3}^{(i)}, \ldots, d_{r(D_r-1)}^{(i)}$, $i = 0,1$ and 2 and D_r which corresponds to the number of bits of the sequence *r*.

The sequence $d_{r0}^{(0)}, d_{r1}^{(0)}, d_{r2}^{(0)}, d_{r3}^{(0)}, \ldots, d_{r(D_r-1)}^{(0)}$ is equal to the sequence $c_{r0}, c_{r1}, c_{r2}, c_{r3}, \ldots, c_{r(K_r-1)}$.

The input of the first encoder is the data structure $c_{r0}, c_{r1}, c_{r2}, c_{r3}, \ldots, c_{r(K_r-1)}$. The sequence $d_{r0}^{(1)}, d_{r1}^{(1)}, d_{r2}^{(1)}, d_{r3}^{(1)}, \ldots, d_{r(D_r-1)}^{(1)}$ is generated at the output of the first encoder.

The input of the second encoder is the output of the QPP interleaving $c'_0, c'_1, \ldots, c'_{K-1}$. The sequence $d_{r0}^{(2)}, d_{r1}^{(2)}, d_{r2}^{(2)}, d_{r3}^{(2)}, \ldots, d_{r(D_r-1)}^{(2)}$ is generated at the output of the encoder.

7.5.4. *Rate matching*

The three sequences issued from the turbo code are interleaved and then stored in a circular memory.

The propagation conditions determine the scheme of the retained coding. The rate matching determines the sequence with the desired coding rate.

The bits of the circular memory are selected or punctured to constitute the output sequence $e_{r0}, e_{r1}, e_{r2}, e_{r3}, \ldots, e_{r(E_r-1)}$, where E_r corresponds to the number of bits of the sequence.

The Hybrid Automatic Repeat reQuest (HARQ) is integrated into the rate matching process.

The output sequence of the circular memory is transmitted from a starting position given by the Redundancy Version (RV).

Different redundancy versions allow for the retransmission of the selected data.

For the Chase Combining mechanism, retransmission contains the same sequence. The error correction can be improved by increasing the signal-to-noise ratio.

For the incremental redundancy mechanism, retransmission contains different sequences. The error correction can be improved by increasing the number of redundancy bits.

7.5.5. *Concatenation*

The different sequences $e_{r0}, e_{r1}, e_{r2}, e_{r3}, \ldots, e_{r(E_r-1)}$, with $r = 0, \ldots, C-1$, where C corresponds to the number of segments, are concatenated to constitute the sequence $f_0, f_1, f_2, f_3, \ldots, f_{G-1}$, where G corresponds to the number of bits of the sequence.

7.5.6. *Scrambling*

The bit sequence $b^{(q)}(0), \ldots, b^{(q)}(M_{\text{bit}}^{(q)}-1)$, $M_{\text{bit}}^{(q)}$ which designates the number of bits for the code word q, is scrambled by the sequence $c^q(i)$; the code word corresponds to an encoded transport block.

This results in the bit sequence $\tilde{b}^{(q)}(0), \ldots, \tilde{b}^{(q)}(M_{\text{bit}}^{(q)}-1)$, with $\tilde{b}^q(i) = \left(b^q(i) + c^q(i)\right) \bmod 2$.

The scrambling sequence is built by means of a pseudo-random sequence that uses a 31-bit polynomial.

The initial value c_{init} of the scrambling sequence is calculated with the following formula:

$$c_{\text{init}} = n_{\text{RNTI}} \cdot 2^{14} + q \cdot 2^{13} + \lfloor n_s/2 \rfloor \cdot 2^9 + N_{\text{ID}}^{\text{cell}}$$

– n_{RNTI} corresponds to the Radio Network Temporary Identifier (RNTI), allocated to the mobile during the connection;

– q is the code word index, $q \in \{0,1\}$, $q = 0$ where a single code word is used;

– n_s corresponds to the time slot number of the time frame;

– $N_{\text{ID}}^{\text{cell}}$ corresponds to the Physical-layer Cell Identity (PCI).

7.5.7. *Modulation*

The propagation conditions determine which modulation scheme is retained: Quadrature Phase-Shift Keying (QPSK), 16-Quadrature Amplitude Modulation (16-QAM) or 64-QAM.

The symbol sequence $d^{(q)}(0),...,d^{(q)}(M_{\text{symb}}^{(q)}-1)$, obtained from the modulation, is generated from the bit sequence $\tilde{b}^{(q)}(0),...,\tilde{b}^{(q)}(M_{\text{bit}}^{(q)}-1)$.

7.5.8. *Mapping on the spatial layers*

The set of symbols $d^{(q)}(0),...,d^{(q)}(M_{\text{symb}}^{(q)}-1)$ is mapped on the set $x^{(0)}(i)$ ... $x^{(\upsilon-1)}(i)$, with $i=0,1,...,M_{\text{symb}}^{\text{layer}}-1$, where υ corresponds to the number of spatial layers and $M_{\text{symb}}^{\text{layer}}$ corresponds to the number of symbols per spatial layer.

For Release 8 and 9, when spatial multiplexing is used, the symbols are mapped on one to four spatial layers (Figures 7.12 and 7.13).

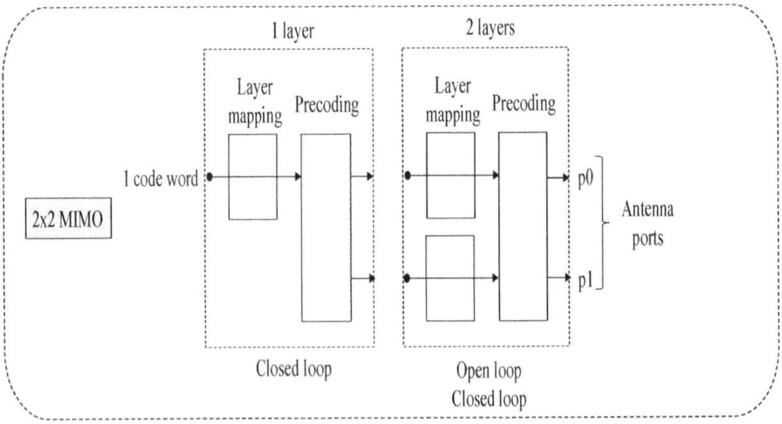

Figure 7.12. *Layer mapping for Release 8 and 9: 2×2 MIMO*

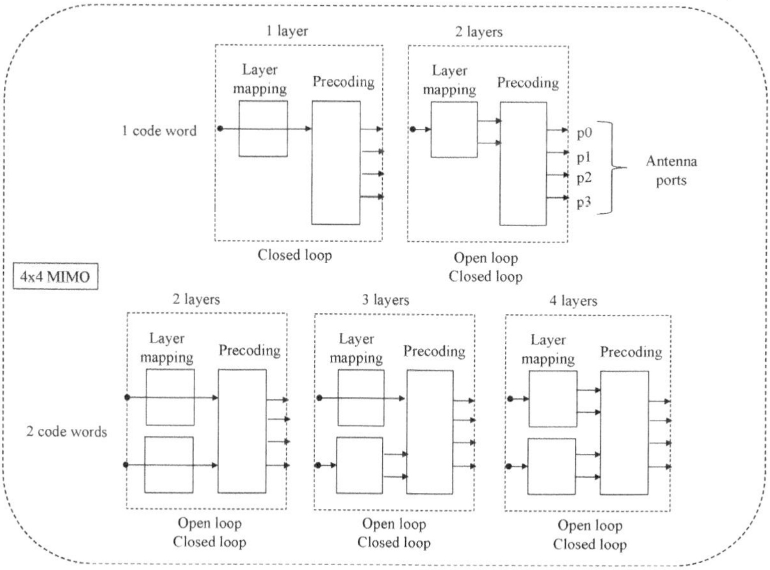

Figure 7.13. *Layer mapping for Release 8 and 9: 4×4 MIMO*

For Release 10, when spatial multiplexing is used, the symbols are mapped on one to eight spatial layers (Figure 7.14).

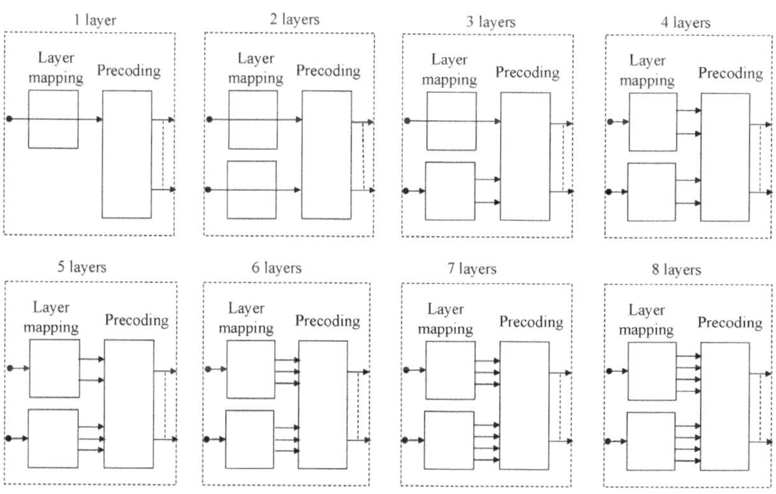

Figure 7.14. *Layer mapping for Release 10*

When using transmit diversity, a single code word is used; the symbols are mapped on two or four layers.

7.5.9. *Precoding*

The precoding is applied to the set of symbols $x^{(0)}(i)$ ... $x^{(\upsilon-1)}(i)$ to generate the sequences $y^{(p)}(i)$ that correspond to the transmission over the antenna port p.

When a single layer is used, no precoding is performed:

– $y^{(p)}(i) = x^{(0)}(i)$, with $p \in \{0,4,5,7,8\}$ corresponding to the antenna port number used for the transmission.

7.5.9.1. *Spatial multiplexing*

For Release 8, the spatial multiplexing, using the Cell-Specific Reference Signal (RS), supports two or four antenna ports, with $p \in \{0,1\}$ or $p \in \{0,1,2,3\}$ (Figures 7.12 and 7.13).

Two configuration modes for the precoding matrix are defined, one with a Cyclic Delay Diversity (CDD), the other without the CDD.

The method with CDD that corresponds to an open-loop spatial multiplexing consists of introducing a different delay for each antenna port.

For the method without CDD, which corresponds to a closed-loop multiplexing, the mobile returns to the evolved Node B (eNB) entity a Precoding Matrix Indicator (PMI) that corresponds to an entry in a dictionary constituted by a set of precoding values.

For Release 10, the spatial multiplexing, using the UE-Specific RS physical signal, supports up to eight antenna ports $p = 7,8,...,\upsilon+6$ (Figure 7.14).

7.5.9.2. *Transmit diversity*

When using two antenna ports (p0 and p1), the PDSCH physical channel is transmitted through Space Frequency Block Coding (SFBC), a form of transmit diversity.

The symbols are transmitted using the following precoding:

– antenna Port (0) $y^{(0)}(2i) = x^{(0)}(i)$ $y^{(0)}(2i+1) = x^{(1)}(i)$;

– antenna Port (1) $y^{(1)}(2i) = -(x^{(1)}(i))^*$ $y^{(1)}(2i+1) = (x^{(0)}(i))^*$.

When using four antenna ports (p0 to p3), the PDSCH physical channel is transmitted through SFBC/FSTD (Frequency Shift Transmit Diversity).

The symbols are transmitted using the following precoding:

– antenna Port (0) $y^{(0)}(4i) = x^{(0)}(i)$ $y^{(0)}(4i+1) = x^{(1)}(i)$;

– antenna Port (1) $y^{(1)}(4i+2) = x^{(2)}(i)$ $y^{(1)}(4i+3) = x^{(3)}(i)$;

– antenna Port (2) $y^{(2)}(4i) = -(x^{(1)}(i))^*$ $y^{(2)}(4i+1) = (x^{(0)}(i))^*$;

– antenna Port (3) $y^{(3)}(4i+2) = -(x^{(3)}(i))^*$ $y^{(3)}(4i+3) = (x^{(2)}(i))^*$.

7.5.10. *Mapping on the resource elements*

The sequence of symbols $y^{(p)}(0),...,y^{(p)}(M_{\text{symb}}-1)$ is mapped on the resource elements located by the index pair (k,l):

– the index k corresponds to the resource element number in the frequency domain;

– the index l corresponds to the OFDM symbol number in the time domain.

The mapping on the resource elements is performed by increasing in the first place, the index k for the lower value of the index l.

The mapping on resource elements continues by increasing the index l by one unit and continuing with the previous operation.

The mapping must not use resource elements allocated to the physical signals and physical channels allocated to the control.

Figure 7.15 shows an example of the mapping on Resource Blocks (RB) allocated to the PDSCH physical channel, in frames that do not contain the

Primary Synchronization Signal (PSS), the Secondary Synchronization Signal (SSS) and the Physical Broadcast Channel (PBCH).

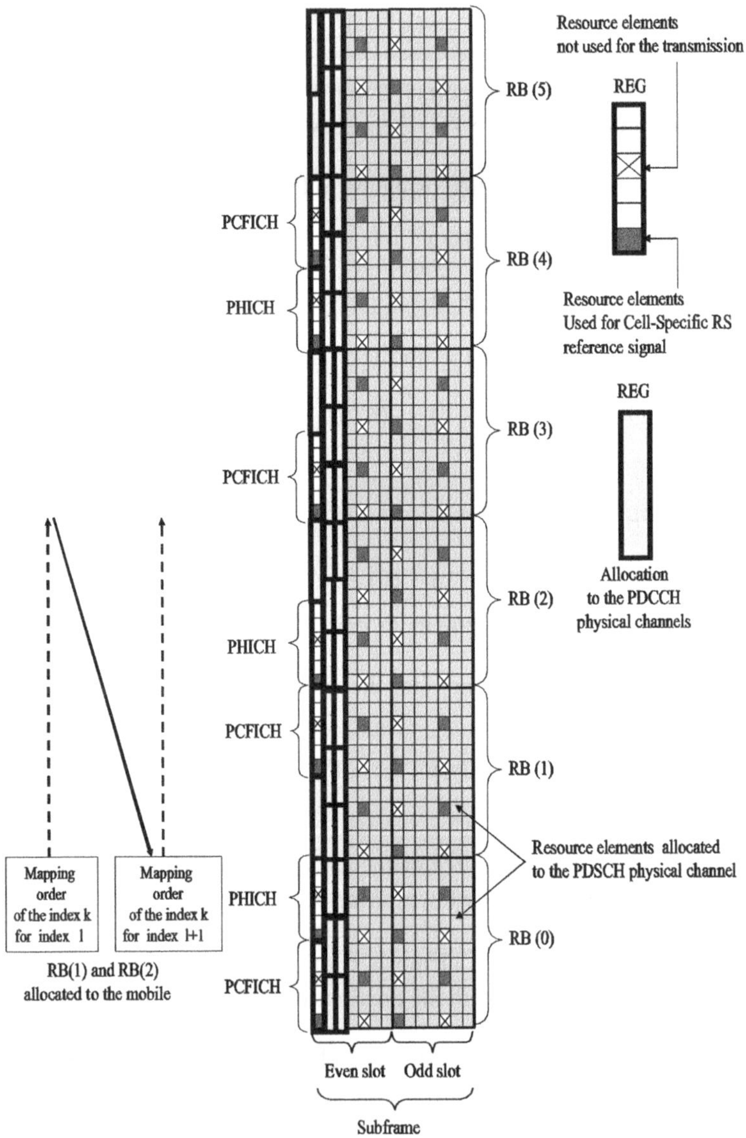

Figure 7.15. *Mapping of the PDSCH physical channel*

7.5.11. *Resource allocation*

7.5.11.1. *Type-0 resource allocation*

The type-0 resource allocation is signaled in the Physical Downlink Control Channel (PDCCH), by means of the Downlink Control Information (DCI) in format 1, 2, 2A, 2B and 2C.

The type-0 resource allocation includes a bit mapping that specifies the Resource Block Groups (RBGs) allocated to the mobile.

The size (*P*) of the resource block group depends on the bandwidth of the radio channel.

The number (N_{RBG}) of resource block groups is equal to $\lceil N_{\text{RB}}^{\text{DL}} / P \rceil$ and corresponds to the bit mapping size (Table 7.18).

$N_{\text{RB}}^{\text{DL}}$ corresponds to the bandwidth of the radio channel, expressed as the number of resource blocks.

Radio channel	1,4-MHz	3-MHz	5-MHz	10-MHz	15-MHz	20-MHz
$N_{\text{RB}}^{\text{DL}}$	6	15	25	50	75	100
P	1	2	2	3	4	4
N_{RBG}	6	8	13	17	19	25

Table 7.18. *Features of the type-0 resource allocation*

Figure 7.16 depicts the allocation resource principles for a radio channel with a 3-MHz bandwidth.

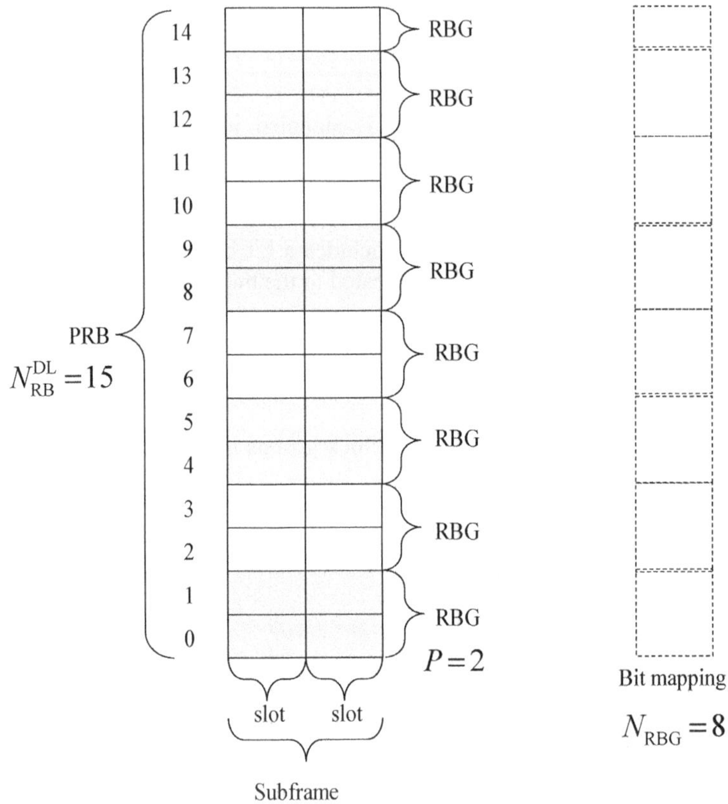

Figure 7.16. *Type-0 resource allocation*

7.5.11.2. *Type-1 resource allocation*

The type-1 resource allocation is signaled in the Physical Downlink Control Channel (PDCCH), by means of the DCI information in format 1, 2, 2A, 2B and 2C.

The type-1 resource allocation cannot be applied in the radio channel with a 1.4-MHz bandwidth.

A subset contains the resource block groups separated by (P) which designates the size of a resource block group.

The number of subsets is therefore equal to (P).

The number of bits to encode the subset of resource block groups is equal to $\lceil \log_2(P) \rceil$ (Table 7.19).

A bit mapping is used to address the resource block group in a subset.

The size of the bit mapping N_{RB}^{TYPE1} is obtained with the following formula:

$$N_{RB}^{TYPE1} = \lceil N_{RB}^{DL} / P \rceil - \lceil \log_2(P) \rceil - 1$$

The size N_{RB}^{TYPE1} does not allow addressing all the resource block groups in a subset.

A shift flag, encoded into a bit, allows covering all the resource block groups:

– bit set to ZERO: the least significant bit corresponds to the resource block with the lowest frequency;

– bit set ONE: the most significant bit corresponds to the resource block with the highest frequency.

Radio channel	1,4-MHz	3-MHz	5-MHz	10-MHz	15-MHz	20-MHz
N_{RB}^{DL}	N.A.	15	25	50	75	100
$\lceil \log_2(P) \rceil$	N.A.	1	1	2	2	2
N_{RB}^{TYPE1}	N.A.	6	11	14	16	22
Flag	N.A.	1	1	1	1	1
Number of bits	N.A.	8	13	17	19	25

N.A. not available

Table 7.19. *Features of the type-1 resource allocation*

Figure 7.17 depicts the allocation resource principles for a radio channel with a 3-MHz bandwidth.

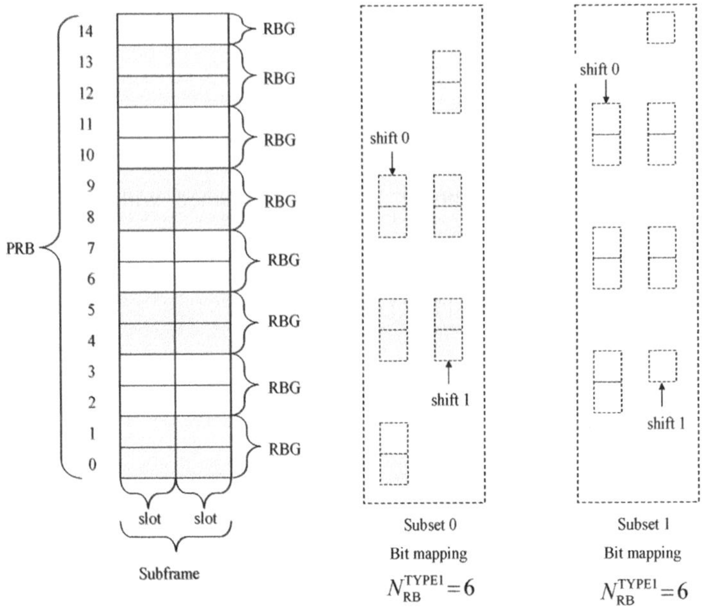

Figure 7.17. *Type-1 resource allocation*

7.5.11.3. *Type-2 resource allocation*

The type-2 resource allocation is signaled in the Physical Downlink Control Channel (PDCCH), by means of the DCI information in format 1A, 1B, 1C and 1D.

The type-2 allocates a set of Virtual Resource Blocks (VRB), mapped over the Physical Resource Blocks PRB.

Localized mapping occurs when adjacent virtual resource blocks correspond to physical resource blocks.

Distributed mapping occurs when adjacent virtual resource blocks correspond to physical resource blocks distributed over the bandwidth of the radio channel.

The mapping can be localized or distributed when the resource allocation uses the DCI information with format 1A, 1B and 1D.

The mapping can only be localized when the resource allocation uses the DCI information with format 1C.

7.5.11.3.1. Localized mapping

The resource allocation can range from 1 to the complete set of physical resource blocks.

The resource allocation uses the parameter Resource Indication Value (RIV) that corresponds to a start resource block RB_{start} and to a number L_{CRBs} of adjacent virtual resource blocks.

$$RIV = N_{RB}^{DL}(L_{CRBs} - 1) + RB_{start} \text{ if } (L_{CRBs} - 1) \leq \lfloor N_{RB}^{DL} / 2 \rfloor$$

$$RIV = N_{RB}^{DL}(N_{RB}^{DL} - L_{CRBs} + 1) + (N_{RB}^{DL} - 1 - RB_{start}) \text{ in the opposite case.}$$

$N_{\mathrm{RB}}^{\mathrm{DL}}$ corresponds to the bandwidth of the radio channel, expressed as the number of resource blocks.

The value of RB_{start} ranges from 0 to $N_{\mathrm{RB}}^{\mathrm{DL}} - 1$.

Figure 7.18 shows the resource block allocation, in a radio channel with a 3-MHz bandwidth, for which the RIV parameter values are equal to 50 (case #1) and to 100 (case #2).

$N_{\mathrm{RB}}^{\mathrm{DL}} = 15$

Case #1

$$L_{CRBs} = \lfloor RIV / N_{RB}^{DL} \rfloor + 1 = 4,\ (L_{CRBs} - 1) \leq \lfloor N_{RB}^{DL} / 2 \rfloor \text{ is verified.}$$

$$RB_{start} = RIV - N_{RB}^{DL}(L_{CRBs} - 1) = 5$$

Cas #2

$$L_{CRBs} = N_{RB}^{DL} - \lfloor RIV / N_{RB}^{DL} \rfloor + 1 = 10,\ (L_{CRBs} - 1) > \lfloor N_{RB}^{DL} / 2 \rfloor \text{ is verified}$$

$$RB_{start} == N_{RB}^{DL}(N_{RB}^{DL} - L_{CRBs} + 2) - RIV - 1) = 4$$

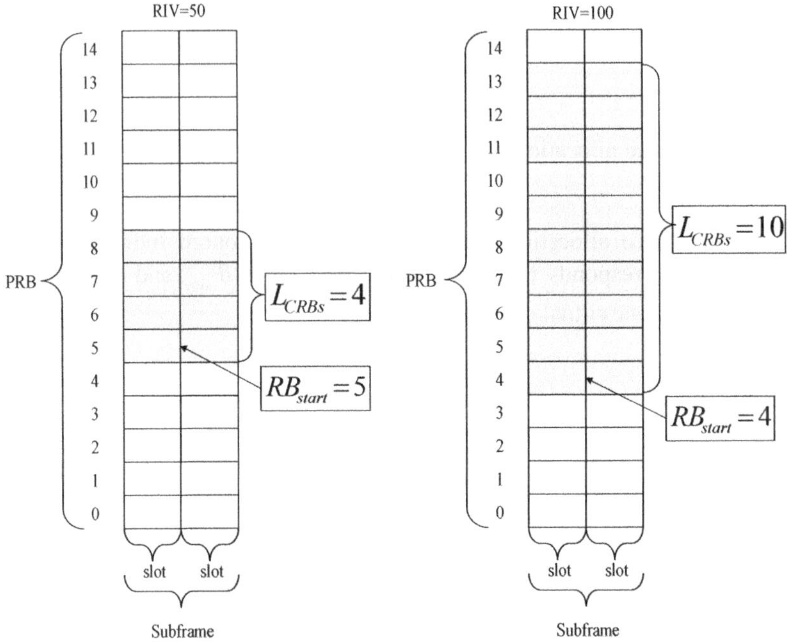

Figure 7.18. *Type-2 resource allocation: localized mapping*

7.5.11.3.2. Distributed mapping

The distributed mapping allows us to spread the resource blocks over the bandwidth of the radio channel, with a shift N_{gap} between the two time slots of the subframe, where N_{gap} can take two values $N_{gap,1}$ or $N_{gap,2}$, each value corresponding to a specific distribution (Table 7.20).

Radio channel	1,4-MHz	3-MHz	5-MHz	10-MHz	15-MHz	20-MHz
$N_{gap,1}$	3	8	12	27	32	48
$N_{gap,2}$	N.A.	N.A.	N.A.	9	16	16

N.A. not available

Table 7.20. *Value of the shifts* $N_{gap,1}$ *and* $N_{gap,2}$

The variation range for the allocation of the virtual resource blocks is determined by the format of the DCI information, the type of RNTI (P, RA, SI, C, TC, SPS) and the shift N_{gap} (Table 7.21).

Shift	Radio channel	1, 4-MHz	3-MHz	5-MHz	10-MHz	15-MHz	20-MHz
$N_{gap,1}$	DCI 1A (P, RA, SI)	1–6	1–14	1–24	1–46	1–64	1–96
	DCI 1A (C, TC, SPS), 1B, 1D	1–6	1–14	1–24	1–16	1–16	1–16
	DCI 1C	2–6	2–14	2–24	4–44	4–64	4–96
$N_{gap,2}$	DCI 1A (P, RA, SI)	N.A.	N.A.	N.A.	1–36	1–64	1–96
	DCI 1A (C, TC, SPS), 1B, 1D	N.A.	N.A.	N.A.	1–16	1–16	1–16
	DCI 1C	N.A.	N.A.	N.A.	4–36	4–64	4–96

N.A. not available

Table 7.21. *Variation ranges for the allocated VRB*

For the DCI information with format 1C, the number of the virtual resource blocks and of the parameter RB_{start} have particular values:

– the value is a multiple of 2 for the radio channel with a 1.4-MHz, 3-MHz or 5-MHz bandwidth;

– the value is a multiple of 4 for the radio channel with a 10-MHz, 15-MHz or 20-MHz bandwidth.

If the shift $N_{gap,1}$ is applied, the mapping between the virtual resource blocks and the physical resource blocks uses an interleaving table that systematically has 4 columns.

The interleaving table is written row by row by the numbers of the virtual resource groups and is read column by column for the allocation of the physical resource blocks:

– the virtual resource blocks are placed at the end of the second and fourth columns;

– the physical resource blocks issued from the first and the second column are aligned with the bottom of the radio channel;

– the physical resource blocks issued from the third and the fourth column are aligned with the top of the radio channel;

– the physical resource blocks not in use, are available for the localized mapping.

Figure 7.19 depicts the allocation of the physical resource blocks in the radio channel with a 3-MHz bandwidth for the following conditions:

$$RIV = 50$$

$$L_{CRBs} = 4$$

$$RB_{start} = 5$$

$$N_{gap,1} = 8$$

Maximum value of allocated virtual resource blocks: 14

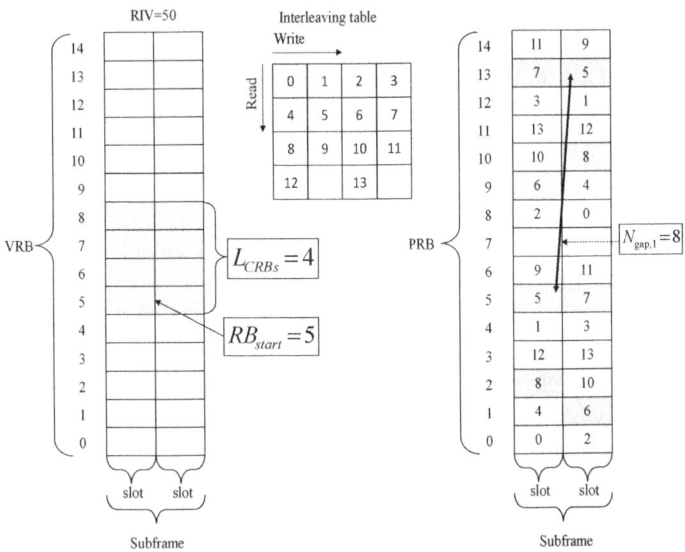

Figure 7.19. *Type-2 resource allocation: distributed mapping*

If the shift $N_{gap,2}$ is applied, the mapping between the virtual resource blocks and the physical resource blocks uses two interleaving tables; both tables systematically have 4 columns.

Each interleaving table is written, row by row, by the numbers of the virtual resource blocks and is read, column by column, for the allocation of resource blocks:

– free virtual resource blocks are placed at the end of the second and fourth columns of each table;

– the physical resource blocks issued from each table are aligned with the bottom of the radio channel;

– the physical resource blocks not used are localized at the top of the radio channel.

7.6. PMCH physical channel

The Physical Multicast Channel (PMCH) transmits the Multicast Channel (MCH).

The MCH transport channel contains traffic data related to Multimedia Broadcast Multicast Service (MBMS), and RRC control data related to the configuration of the MBMS Single Frequency Network (MBSFN) area and to the counting requests.

The processing associated with the PMCH physical channel is summarized in Figure 7.20.

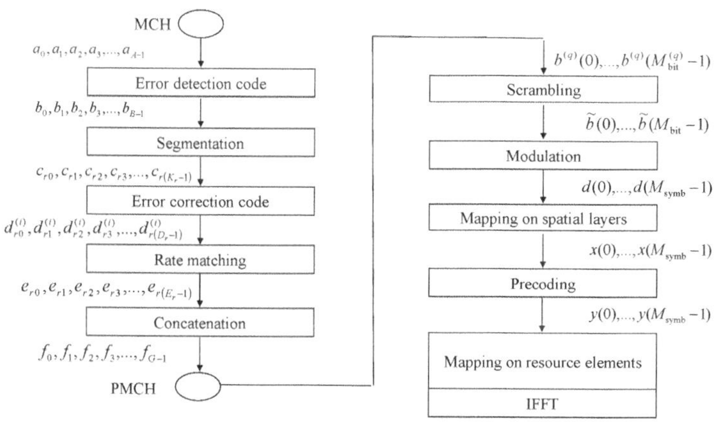

Figure 7.20. *Processing associated with the PMCH physical channel*

7.6.1. *Error detection codes*

The error detection is obtained from a Cyclic Redundancy Check (CRC) code.

The CRC structure is the remainder when the transport block $a_0, a_1, a_2, a_3, ..., a_{A-1}$ is divided by a generator polynomial, whose remainder $p_0, p_1, p_2, p_3, ..., p_{L-1}$ constitutes the bits of cyclic redundancy.

The concatenation of the transport block $a_0, a_1, a_2, a_3, ..., a_{A-1}$ and of the 24 bits of the cyclic redundancy $p_0, p_1, p_2, p_3, ..., p_{L-1}$ constitute the input structure $b_0, b_1, b_2, b_3, ..., b_{B-1}$ of the segmentation.

7.6.2. *Segmentation*

The maximum size of the block processed by the error correction code is equal to 6144 bits.

If the size of the block $b_0, b_1, b_2, b_3, ..., b_{B-1}$ is above this value, it must be segmented, and every segment must have its own CRC error detection code, whose length is equal to 24 bits in order to constitute the sequence $c_{r0}, c_{r1}, c_{r2}, c_{r3}, ..., c_{r(K_r-1)}$, where r and Kr correspond respectively to the number of segment and to the number of bits of the segment.

7.6.3. *Error correction code*

When segmentation is performed over the sequence $b_0, b_1, b_2, b_3, ..., b_{B-1}$, the error correction mechanism is applied to every segment.

The error correction process is performed by a turbo code made up by a Parallel Concatenated Convolutional Code (PDCCC) and by a Quadratic Permutation Polynomial (QPP) interleaving (Figure 7.11).

The error correction code produces three sequences $d_{r0}^{(i)}, d_{r1}^{(i)}, d_{r2}^{(i)}, d_{r3}^{(i)}, ..., d_{r(D_r-1)}^{(i)}$, with $i = 0,1$ and 2 where D_r corresponds to the number of bits of the sequence.

The sequence $d_{r0}^{(0)}, d_{r1}^{(0)}, d_{r2}^{(0)}, d_{r3}^{(0)}, ..., d_{r(D_r-1)}^{(0)}$ is equal to sequence $c_{r0}, c_{r1}, c_{r2}, c_{r3}, ..., c_{r(K_r-1)}$.

The input of the first encoder is the data structure $c_{r0}, c_{r1}, c_{r2}, c_{r3}, ..., c_{r(K_r-1)}$. The sequence $d_{r0}^{(1)}, d_{r1}^{(1)}, d_{r2}^{(1)}, d_{r3}^{(1)}, ..., d_{r(D_r-1)}^{(1)}$ is produced at the output of the first encoder.

The input of the second encoder is the output of the QPP interleaving $c'_0, c'_1, ..., c'_{K-1}$. The sequence $d_{r0}^{(2)}, d_{r1}^{(2)}, d_{r2}^{(2)}, d_{r3}^{(2)}, ..., d_{r(D_r-1)}^{(2)}$ is produced at the output of the second encoder.

7.6.4. *Rate matching*

The three sequences issued from the turbo code are interleaved and then stored in a circular memory.

The bits of the circular memory are selected or punctured to constitute the output sequence $e_{r0}, e_{r1}, e_{r2}, e_{r3}, ..., e_{r(E_r-1)}$ where E_r corresponds to the number of the bits of the sequence.

7.6.5. *Concatenation*

All different sequences $e_{r0}, e_{r1}, e_{r2}, e_{r3}, ..., e_{r(E_r-1)}$, with $r = 0, ..., C-1$, where C corresponds to the number of segments, are concatenated to constitute the sequence $f_0, f_1, f_2, f_3, ..., f_{G-1}$, where G corresponds to the number of bits of the sequence.

7.6.6. *Scrambling*

The sequence of bits $b(0), ..., b(M_{\text{bit}} - 1)$, M_{bit} that designates the number of bits, is scrambled by the sequence $c(i)$.

This generates the sequence of bits $\tilde{b}(0), ..., \tilde{b}(M_{\text{bit}} - 1)$, with $\tilde{b}(i) = \left(b(i) + c(i)\right) \bmod 2$.

The scrambling sequence is obtained with a pseudo-random sequence that used a 31-bit polynomial.

The initial value c_{init} of the scrambling sequence is calculated with the following formula:

$$c_{\text{init}} = n_{\text{RNTI}} \cdot 2^{14} + \lfloor n_s/2 \rfloor \cdot 2^9 + N_{\text{ID}}^{\text{cell}}$$

– n_{RNTI} corresponds to the Radio Network Temporary Identifier (RNTI), allocated to the mobile during the connection;

– n_s corresponds to the time slot number of the time frame;

– $N_{\text{ID}}^{\text{cell}}$ corresponds to the Physical-layer Cell Identity (PCI).

7.6.7. *Modulation*

The sequence of the symbols $d(0),...,d(M_{\text{symb}}-1)$, issued from the Quadrature Phase-Shift Keying (QPSK) modulation, the 16-Quadrature Amplitude Modulation (16-QAM) or the 64-QAM modulation, is obtained with the sequence of bits $b(0),...,b(M_{\text{bit}}-1)$.

7.6.8. *Mapping on the spatial layers*

The PMCH physical signal uses a single spatial layer. The set of symbols $d(0),...,d(M_{\text{symb}}-1)$ is mapped in the set $x(0),...,x(M_{\text{symb}}-1)$.

7.6.9. *Precoding*

When a single spatial layer is used, no precoding is performed. The set of symbols $x(0),...,x(M_{\text{symb}}-1)$ is mapped in the set $y(0),...,y(M_{\text{symb}}-1)$.

7.6.10. *Mapping on the resource elements*

The PMCH physical signal occupies all the Resource Blocks (RB) in the frequency domain, for the duration of a subframe.

The PMCH physical signal uses an extended cyclic prefix and a 15 kHz or 7.5 kHz step between subcarriers.

The control area allocated to the Physical Control Format Indicator Channel (PCFICH), the Physical HARQ Indicator Channel (PHICH) and the Physical Downlink Control Channel (PDCCH), is limited to two Orthogonal Frequency Division Multiplexing (OFDM) symbols.

The PDCCH physical channel is solely used to supply the resources to the mobile for the uplink direction, over the Physical Uplink Control Channel (PUSCH).

The control area uses the cyclic prefix defined in the cell for the non-MBSFN subframes.

8

Uplink Physical Signals

8.1. DM-RS physical channel

The Demodulation Reference Signal (DM-RS) associated with the Physical Uplink Shared Channel (PUSCH) is used for the assessment and the synchronization of the PUSCH physical channel.

The DM-RS physical signal associated with the Physical Uplink Control Channel (PUCCH) is used for the assessment and the synchronization of the PUCCH physical channel.

8.1.1. *DM-RS physical signal associated with PUSCH physical channel*

8.1.1.1. *Sequence generation*

For Release 8 and 9, the transmission of the PUSCH physical channel uses a single antenna port.

The DM-RS physical signal corresponds to the following formula:

$$r_{\text{PUSCH}} = r_{u,v}^{(\alpha)}(n) = e^{j\alpha n)}.\overline{r}_{u,v}(n)$$

– α corresponds to the cyclic shift applied to the base sequence $\overline{r}_{u,v}(n)$;

– α can take 12 values of cyclic shift;

– u corresponds to the group number of the base sequence;

– u can take 30 values, from 0 to 29;

– v corresponds to the index of the base sequence in the group;

– v can take two values, 0 or 1, if the size of the DM-RS physical signal is greater than or equal to six Resource Blocks (RB) in the frequency domain;

– v otherwise it takes a single value, 0;

– n takes the value between 0 and $M_{sc}^{RS}-1$, $M_{sc}^{RS}=mN_{sc}^{RB}$ which is the length of the reference signal;

– m corresponds to the number of resource blocks allocated to the mobile;

– N_{sc}^{RB} corresponds to the number of subcarriers per resource block in the frequency domain.

If the DM-RS physical signal has a size greater than or equal to $3N_{sc}^{RB}$, the base sequence is generated from a Zadoff-Chu sequence of N_{ZC}^{RS} length, as N_{ZC}^{RS} is the largest prime number less than M_{sc}^{RS}.

If the DM-RS physical signal is less than $3N_{sc}^{RB}$, the base sequence is generated from the formula: $\bar{r}_{u,v}(n)=e^{j\phi(n)\pi/4}, \quad 0\leq n\leq M_{sc}^{RS}-1$.

The value of $\phi(n)$ depends on the group number of the base sequence u.

The group number of the base sequence is determined from the following formula:

$$u=\left(f_{gh}(n_s)+f_{ss}\right)\bmod 30.$$

– $f_{gh}(n_s)$ is a random number between 0 and 29 if the group hopping is activated. Otherwise, it takes the value equal to 0;

– $f_{gh}(n_s)$ depends on the value of the time slot n_s and of the value N_{ID}^{cell} of the Physical-layer Cell Identity (PCI);

– $f_{ss} = \left(N_{ID}^{cell} + \Delta_{ss}\right) \bmod 30$, the $\Delta_{ss} \in \{0,1,...,29\}$ value is shown in the System Information Block 2 (SIB2) message.

The index of the base sequence v takes a value equal to 1 if the group hopping is deactivated and if the sequence hopping is activated in the SIB2 system information message.

The index of the base sequence $v=1$ determines the use of a sequence, obtained from a pseudo-random sequence whose initial value depends on the value of the parameters N_{ID}^{cell} and f_{ss}.

For Release 10, the transmission of the PUSCH physical channel can use four antenna ports.

Using the Orthogonal Cover Code (OCC) facilitates the spatial multiplexing of several mobiles for multi-user the Multiple Input Multiple Output (MIMO) transmission mode.

The DM-RS physical signal is generated from the following formula:

$$r_{\text{PUSCH}} = w^{(\lambda)}(m) r_{u,v}^{(\alpha_\lambda)}(n).$$

– λ corresponds to the antenna port number;

– $w^{(\lambda)}$ is the value of the OCC orthogonal sequence;

– m takes the value 0 for the first time slot or 1 for the second time slot;

– α_λ corresponds to the cyclic shift applied to the base sequence, depending on the antenna port number λ.

8.1.1.2. *Mapping on the resource elements*

The DM-RS physical signal is mapped on the fourth Orthogonal Frequency Division Multiplexing (OFDM) symbol in the case of a normal cyclic prefix or on the third OFDM symbol in case of an extended cyclic prefix (Figure 8.1).

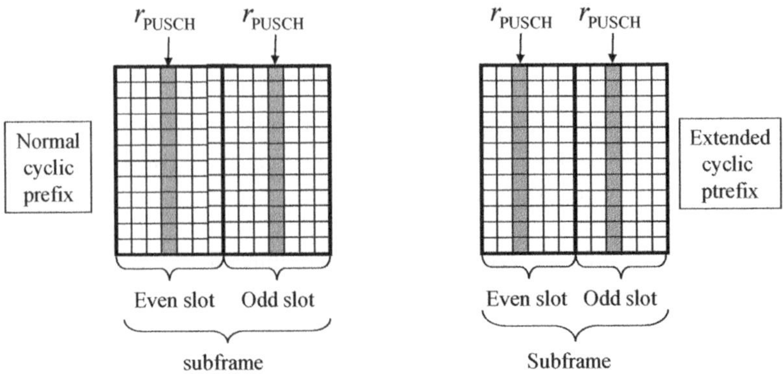

Figure 8.1. *Mapping of the DM-RS physical signal associated with the PUSCH physical channel*

8.1.2. *DM-RS physical signal associated with the PUCCH physical channel*

8.1.2.1. *Sequence generation*

For Release 8 and 9, the DM-RS physical signal corresponds to the following formula:

$$r_{\text{PUCCH}} = \overline{w}(m)z(m)r_{u,v}^{(\alpha)}(n)$$

– $\overline{w}(m)$ is an orthogonal sequence only applicable to formats 1, 1a and 1b of the PUCCH physical channel, allowing the multiplexing by the code of several mobiles;

– $z(m)$ takes the value equal to 1 by default. For the second OFDM symbol of format 2a and 2b, $z(m)$ is equal to the eleventh symbol of the PUCCH physical channel;

– (m) references the OFDM symbol number allocated to the DM-RS physical signal, for example (m) varies from 0 to 2 for formats 1, 1a and 1b;

– n takes the value between 0 and 11. Each value corresponds to a subcarrier of the resource block allocated to the PUCCH physical channel.

The parameter calculation $f_{ss} = \left(N_{ID}^{cell}\right) \bmod 30$ is carried out with a shift Δ_{ss} of zero.

v only takes the value 0.

The base sequence is generated from the formula:

$$\bar{r}_{u,v}(n) = e^{j\phi(n)\pi/4}.$$

The value of $\phi(n)$ depends on the group number of the base sequence u.

For Release 10, the generated DM-RS physical signal is of the form:

$$\frac{1}{\sqrt{P}}\bar{w}^{(\tilde{p})}(m)z(m)r_{u,v}^{(\alpha_{\tilde{p}})}(n)$$

P corresponds to the number of antenna ports. The factor $\frac{1}{\sqrt{P}}$ was introduced in order to keep the power of the DM-RS physical signal constant, independently from the number of antenna ports.

$\tilde{p}$ is the index of the antenna port number.

The values of the orthogonal sequence $\bar{w}^{(\tilde{p})}$ and of the cyclic shift $\alpha_{\tilde{p}}$ of the sequence $r_{u,v}^{(\alpha_{\tilde{p}})}(n)$ depends on the number of the antenna port index $\tilde{p}$.

8.1.2.2. *Mapping on the resource elements*

For the formats 1, 1a and 1b, the DM-RS physical signal is mapped on the third, the fourth and the fifth OFDM symbol in case of a normal cyclic prefix or on the third and the fourth OFDM symbols in the case of an extended cyclic prefix (Figure 8.2).

For the formats 2, 2a, 2b and 3, the DM-RS physical signal is mapped on the second and sixth OFDM symbols in the case of a normal cyclic prefix (Figure 8.2).

For the formats 2 and 3, the DM-RS physical signal is mapped on the fourth OFDM symbol in the case of an extended cyclic prefix (Figure 8.2).

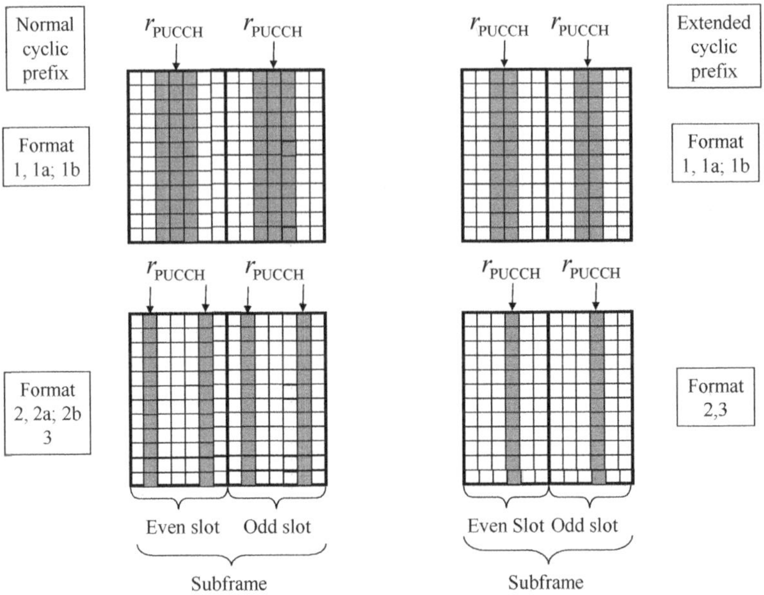

Figure 8.2. *Mapping of the DM-RS physical signal associated with the PUSCH physical channel*

8.2. SRS physical signal

The Sound Reference Signal (SRS) allows the evolved Node B (eNB) entity to measure the quality of the signal for the uplink direction, in a frequency band greater than that allocated to the mobile.

This measure cannot be obtained by the Demodulation Reference Signal (DM-RS) for the latter is associated with the Physical Uplink Shared Channel (PUSCH) or Physical Uplink Control Channel (PUCCH).

The measure carried out by the eNB entity allows it to define the frequency localization of the allocated resource to the mobile for the uplink direction, as well as the Modulation and Coding Scheme (MCS).

For the Time Division Duplex (TDD) mode, the application of the channel reciprocity facilitates the use of the measure of the uplink direction for also qualifying the downlink direction.

8.2.1. *Sequence generation*

For Release 8 and 9, the SRS physical signal is generated from the following formula:

$$r_{\mathrm{SRS}}(n) = r_{u,v}^{(\alpha)}(n) = e^{j\alpha n)}.\bar{r}_{u,v}(n)$$

– α corresponds to the cyclic shift applied to the base sequence $\bar{r}_{u,v}(n)$;

– α can take 8 values of cyclic shift;

– u corresponds to the group number of the base sequence;

– u can take 30 values, from 0 to 29;

– v corresponds to the index of the base sequence in the group;

– v can take two values, 0 or 1, if the size of the SRS physical signal is greater than or equal to six Resource Blocks (RB) in the frequency domain;

– v otherwise it takes a single value, 0;

– n takes the value between 0 and $M_{\mathrm{sc}}^{\mathrm{RS}} - 1$, $M_{\mathrm{sc}}^{\mathrm{RS}}$ which is the length of the reference signal. $M_{\mathrm{sc}}^{\mathrm{RS}} = mN_{\mathrm{sc}}^{\mathrm{RB}}$, $N_{\mathrm{sc}}^{\mathrm{RB}}$ corresponds to the number of subcarriers per resource block in the frequency domain.

If the DM-RS physical signal has a size greater than or equal to $3N_{\mathrm{sc}}^{\mathrm{RB}}$, the base sequence is generated from a Zadoff-Chu sequence of $N_{\mathrm{ZC}}^{\mathrm{RS}}$ length, as $N_{\mathrm{ZC}}^{\mathrm{RS}}$ is the largest prime number less than $M_{\mathrm{sc}}^{\mathrm{RS}}$.

If the SRS physical signal is less than $3N_{\mathrm{sc}}^{\mathrm{RB}}$, the base sequence is generated from the formula:

$$\bar{r}_{u,v}(n) = e^{j\phi(n)\pi/4}, \quad 0 \le n \le M_{\mathrm{sc}}^{\mathrm{RS}} - 1.$$

The value of $\phi(n)$ depends on the group number of the base sequence u.

The group number of the base sequence is determined from the formula:

$$u = \left(f_{\mathrm{gh}}(n_{\mathrm{s}}) + f_{\mathrm{ss}}\right) \bmod 30$$

– $f_{\text{gh}}(n_s)$ is a random number between 0 and 29 if the group hopping is activated. Otherwise, it takes the value equal to 0;

– $f_{\text{gh}}(n_s)$ depends on the value of the time slot n_s and of the value $N_{\text{ID}}^{\text{cell}}$ of the Physical-layer Cell Identity (PCI);

– $f_{\text{ss}} = \left(N_{\text{ID}}^{\text{cell}}\right) \bmod 30$.

The index of the base sequence v takes a value equal to 1 if the group hopping is deactivated and if the sequence hopping is activated in the SIB2 system information message.

The index of the base sequence $v = 1$ determines the use of a sequence, obtained from a pseudo-random sequence whose initial value depends on the value of the parameters $N_{\text{ID}}^{\text{cell}}$ and f_{ss}.

For Release 10, the generated SRS physical signal is of the $r_{\text{SRS}}^{(\tilde{p})}(n) = r_{u,v}^{(\alpha_{\tilde{p}})}(n)$ form. $\tilde{p}$ is the index of the antenna port number.

8.2.2. *Mapping on the resource elements*

The SRS physical signal is transmitted in the last Orthogonal Frequency Division Multiplexing (OFDM) symbol of the subframe.

The cyclic shift facilitates the multiplexing of the SRS physical signals from several mobiles in the Resource Element (RE) of a resource block.

The localization in the SRS signal frequency forms a comb for which the mobile uses one subcarrier out of two.

The bandwidth used for the SRS signal can be reduced, which facilitates a frequency division multiplexing of several mobiles.

The bandwidth is a multiple of four resource blocks in the frequency domain.

The number of resource blocks allocated to the SRS physical signal is indicated by the System Information Block 2 (SIB2) message.

The number of resource blocks used by the mobile and the frequency localization of these blocks are indicated by the Radio Resource Control (RRC) messages.

The localization of the bandwidth used for the SRS physical signal can vary temporally, which facilitates the covering of the radio channel with several transmissions.

In the allocated resources, the SRS physical signal is not transmitted to the PUCCH physical channel which does not need an adjustment mechanism for the quality of the signal.

Figure 8.3 describes the mapping of the SRS physical signal for a radio channel with a bandwidth of 3-MHz:

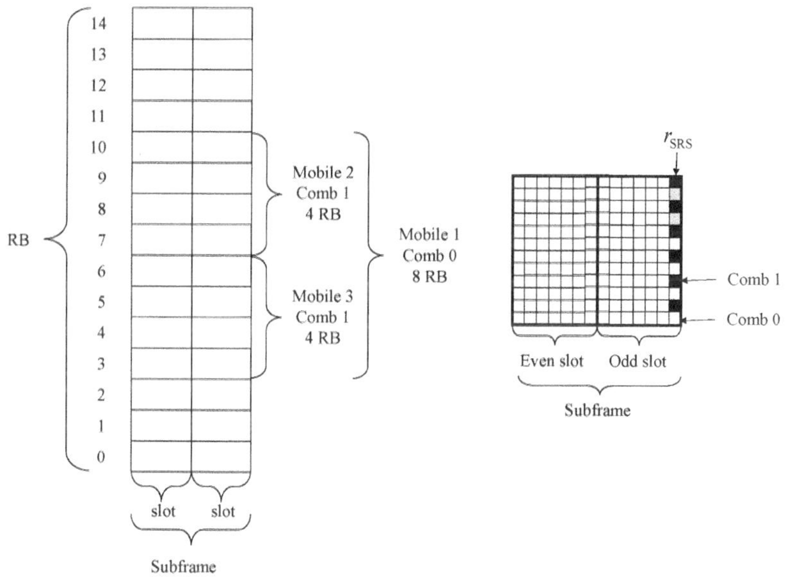

Figure 8.3. *Mapping of the SRS physical signal*

Number of resource blocks allocated to the SRS physical signal: 8

– mobile 1 configuration: 8 RB (3–10), comb 0;

– mobile 2 configuration: 4 RB (7–10), comb 1;

– mobile 3 configuration: 4 RB (3–6), comb 1.

8.2.3. *Configuration of the subframes*

The configuration of the subframes used for the transmission of the SRS physical signal is defined by SIB2 system information and depends on the type of time frame (Tables 8.1 and 8.2).

Table 8.1. *Allocation of the subframe in the SRS physical signal: type-1 time frame*

Number of subframe									
0	1	2	3	4	5	6	7	8	9

Table 8.2. *Allocation of the subframe in the SRS physical signal: type-2 time frame*

For the TDD mode, the SRS physical signal is only transmitted in the subframes allocated to the uplink direction.

Subframes 0 and 5 of the type-2 time frame are always allocated to the downlink direction.

Subframe 1 is always a special frame for which the SRS physical signal is transmitted in the Uplink Pilot Time Slot (UpPTS) field.

Subframe 6 can also contain the UpPTS field according to the configuration of the type-2 time frame.

8.2.4. *Transmission of the SRS physical signal*

The periodicity T_{SRS} of the SRS physical signal transmission can take several values (2, 5, 10, 20, 40, 80, 160 or 320 milliseconds), which facilitates the time division multiplexing of the SRS physical signals from several mobiles belonging to the same cell.

With the exception of 2 ms periodicity for type-1 and type-2 time frames, the SRS physical signal transmission is effective under the following condition:

$$(10 \cdot n_f + k_{\mathrm{SRS}} - T_{offset}) \bmod T_{\mathrm{SRS}} = 0$$

– n_f corresponds to the number of the subframe;

– k_{SRS} takes a value which depends on the number of the subframe and on the length of the UpPTS field;

– T_{offset} is a function of the index I_{SRS} transmitted by the RRC message (Tables 8.3 and 8.4).

In case the periodicity for the type-2 time frame is equal to 2 ms, the SRS physical signal transmission is effective under the following condition:

$$(k_{\mathrm{SRS}} - T_{offset}) \bmod 5 = 0$$

Index I_{SRS}	Periodicity T_{SRS}	Offset T_{offset}
0–1	2	I_{SRS}
2–6	5	$I_{\mathrm{SRS}}-2$
7–16	10	$I_{\mathrm{SRS}}-7$
17–36	20	$I_{\mathrm{SRS}}-17$
37–76	40	$I_{\mathrm{SRS}}-37$
77–156	80	$I_{\mathrm{SRS}}-77$
157–316	160	$I_{\mathrm{SRS}}-157$
317–636	320	$I_{\mathrm{SRS}}-317$

Table 8.3. *Transmission parameters of the SRS physical signal: type-1 time frame*

Index I_{SRS}	Periodicity T_{SRS}	Offset T_{offset}
0	2	0.1
1	2	0.2
2	2	1.2
3	2	0.3
4	2	1.3
5	2	0.4
6	2	1.4
7	2	2.3
8	2	2.4
9	2	3.4
10–14	5	I_{SRS}–10
15–24	10	I_{SRS}–15
25–44	20	I_{SRS}–25
45–84	40	I_{SRS}–45
85–164	80	I_{SRS}–85
165–324	160	I_{SRS}–165
325–644	320	I_{SRS}–325

Table 8.4. *Transmission parameters of the SRS physical signal: type-2 time frame*

The SRS signal transmission can either be alone, when the transmission is carried out on a single frame, or periodic, when the transmission takes place during an undetermined period until the eNB entity interrupts it.

If the selection of the transmit antenna is activated, the SRS physical signal is alternately transmitted on each antenna, which allows the eNB entity to choose the transmit antenna for the PUSCH physical channel.

Release 10 defined two types of triggering of the SRS physical signal transmission:

– type-0 triggering takes the characteristics previously defined by Release 8;

– type-1 triggering is initialized by the Downlink Control Information (DCI).

DCI information in formats 0, 4 and 1A trigger the SRS physical signal transmission, for the two types of time frame.

DCI information in formats 2B and 2C trigger the SRS physical signal transmission, for the type-2 time frame.

The index I_{SRS} is limited to a group between 0 and 16 for type-1 time frame and to a group between 0 and 24 for type-2 time frame.

Release 10 supports the transmission of the SRS physical signal on several antenna ports, and in this case, the different SRS physical signals occupy the same resource block and can be differentiated by a combination of comb and cyclic shift.

Release 10 also supports the transmission of the SRS physical signal on several aggregated radio channels.

8.2.5. *Power control*

The SRS physical signal does not have dedicated power control and shares that of the PUSCH physical channel.

Average power P_{SRS}, expressed in dBm, of the SRS physical signal transmitted on an OFDM symbol is provided by the following formula:

$$P_{\text{SRS}}(i) = \min\left\{ P_{\text{CMAX}}(i), 10\log_{10}(M_{\text{SRS}}) + P_{\text{O\_PUSCH}}(j) + \alpha(j)\cdot PL + f(i) \right\}$$

– $P_{\text{CMAX}}(i)$ corresponds to the configured maximum power of the mobile, for the i subframe;

– M_{SRS} corresponds to the bandwidth used by the SRS physical signal, expressed in the number of resource blocks allocated;

– $P_{\text{O\_PUSCH}}(j)$ is an identical parameter to that used for the power control of the PUSCH physical channel;

– $P_{\text{O\_PUSCH}}(j)$ corresponds to the power amount $P_{\text{O\_NOMINAL\_PUSCH}}(j)$ communicated by the SIB2 system information and $P_{\text{O\_UE\_PUSCH}}(j)$ by an RRC message, for the $j=1$ index that corresponds to a dynamic order;

– α corresponds to a parameter communicated by the SIB2 system information, and it can take the values $\alpha \in \{0, 0.4, 0.5, 0.6, 0.7, 0.8, 0.9, 1\}$;

– PL corresponds to an assessment of the propagation loss calculated by the mobile;

– PL is the difference between the value of the power of the reference signal communicated by the SIB2 system information or an RRC message on one hand, and the value of the Reference Signal Received Power (RSRP) for the reference signal on the other hand;

– $f_c(i)$ corresponds to the Transmit Power Control (TPC) communicated by the eNB entity in the DCI information in formats 0, 3, 3A and 4.

Release 10 introduced a power offset $P_{\text{SRS\_OFFSET}}(m)$ communicated in an RRC message:

– $m=0$ corresponds to triggering of type 0 of the SRS physical signal;

– $m=1$ corresponds to triggering of type 1 of the SRS physical signal.

9

Uplink Physical Channels

9.1. PRACH physical channel

The Physical Random Access Channel (PRACH) contains a preamble used by the mobile when it needs to carry out random access.

The random access is the first step of the mobile connection to the evolved Node B (eNB) entity.

The PRACH physical channel contains a signal using an Orthogonal Frequency Division Multiplexing (OFDM) different from other Physical Uplink Control Channel (PUCCH) and Physical Uplink Shared Channel (PUSCH).

9.1.1. *Time structure of the preamble*

The preamble is made of a Cyclic Prefix (CP) in duration T_{CP} and of a sequence of T_{SEQ} in duration.

The duration of the sequence T_{SEQ} determines the energy collected by the eNB entity in reception.

During the random access, the mobile uses the zero value for timing advance.

The guard time keeps the reception of the PRACH physical channel from disturbing the following subframes for a localized mobile at the edge of the eNB entity (Figure 9.1).

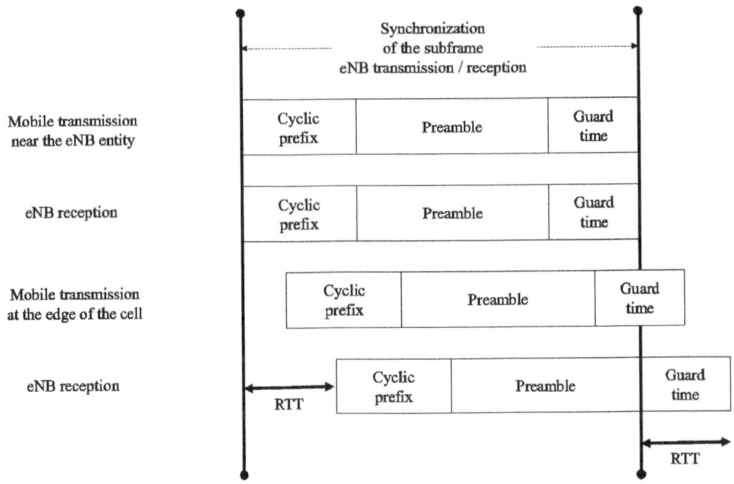

Figure 9.1. *Structure of the PRACH physical channel: guard time*

The guard time is equal to the maximum Round Trip Time (RTT) and determines the maximum value of the radius of the cell:

$$R(\text{km}) = \text{guard time } (\mu s) / 6.6$$

Table 9.1 describes the format of the different preamble.

Format	Cyclic prefix time T_{CP}	Sequence time T_{SEQ}	Guard time	Preamble duration
0	3168×Ts	24576×Ts	2976×Ts	1 subframe
1	21024×Ts	24576×Ts	15840×Ts	2 subframes
2	6240×Ts	2×24576×Ts	6048×Ts	2 subframes
3	21024×Ts	2×24576×Ts	21984×Ts	3 subframes
4	448×Ts	4096×Ts	288×Ts	2 symbols

Note.– Ts = 0.032552 µs

Table 9.1. *Structure of the PRACH physical channel: duration of each field*

For the mobile located at the edge of the cell, the observation window of the eNB entity uses the cyclic prefix to replace the part which is outside the window.

The beginning of the cyclic prefix must take into consideration the RTT duration and the propagation delay spread (Figure 9.2).

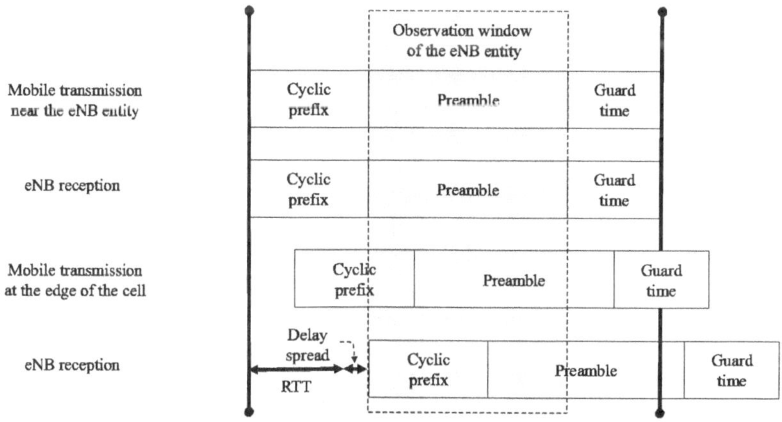

Figure 9.2. *Structure of the PRACH physical channel: observation window*

Delay spread leads to a recovery of the preamble on the following subframe, which is acceptable if the recovery only affects the cyclic prefix of the following subframe.

Format 0 preamble is mapped in a subframe. Maximum size of the cell radius is 14 km.

Format 1 preamble is mapped in two subframes. Maximum size of the cell radius is 75 km.

Format 2 preamble is mapped in two subframes. Maximum size of the cell radius is 28 km.

Format 3 preamble is mapped in three subframes. Maximum size of the cell radius is 108 km.

Format 4 preamble is mapped in the Uplink Pilot Time Slot (UpPTS) field of the special subframe of type-2 time frame. Maximum size of the cell radius is 1.4 km.

Formats 2 and 3 preamble contains two sequences, which allow the eNB entity to obtain two times more energy.

9.1.2. *Frequency structure of the preamble*

The PRACH physical channel uses a bandwidth of 1.08-MHz, which corresponds to 6 Resource Block (RB) in the frequency domain.

For the formats 0 to 3, the subcarriers are separated by 1250 Hz and the PRACH physical channel is made of 864 subcarriers, of which 839 are active.

The guard band of the PRACH physical channel is 16.25 kHz on one side that is 13 subcarriers, and 15 kHz on the other side that is 12 subcarriers.

For format 4, the subcarriers are separated by 7500 Hz and the PRACH physical channel is made of 144 subcarriers, of which 139 are active.

The guard band of the PRACH physical channel is 22.5 kHz on one side that is 3 subcarriers, and 15 kHz on the other side that is 2 subcarriers.

9.1.3. *Location of the preamble*

The subframes allocated to the PRACH physical channel are determined from parameters communicated by System Information Block 2 (SIB2).

9.1.3.1. *Type-1 time frame*

Table 9.2 shows the allocation of the subframes on the PRACH physical channel.

The position $n_{\text{PRB}}^{\text{RA}}$ of the PRACH physical channel is determined from the offset $n_{\text{PRB offset}}^{\text{RA}}$: $n_{\text{PRB}}^{\text{RA}} = n_{\text{PRB offset}}^{\text{RA}}$

9.1.3.2. *Type-2 time frame*

Table 9.3 shows the density of the subframe used by the PRACH physical channel.

The parameter $t_{\text{RA}}^{(0)}$ shows if the PRACH physical channel is transmitted in all the frames, in the even frames or in the odd frames.

The parameter value $t_{\text{RA}}^{(0)}$ depends on the configuration index of the PRACH physical channel.

Index	Format Preamble	Number Frame	Sub Frame	Index	Format Preamble	Number Frame	Sub Frame
0	0	even	1	32	2	even	1
1	0	even	4	33	2	even	4
2	0	even	7	34	2	even	7
3	0	all	1	35	2	all	1
4	0	all	4	36	2	all	4
5	0	all	7	37	2	all	7
6	0	all	1, 6	38	2	all	1, 6
7	0	all	2 ,7	39	2	all	2 ,7
8	0	all	3, 8	40	2	all	3, 8
9	0	all	1, 4, 7	41	2	all	1, 4, 7
10	0	all	2, 5, 8	42	2	all	2, 5, 8
11	0	all	3, 6, 9	43	2	all	3, 6, 9
12	0	all	0, 2, 4, 6, 8	44	2	all	0, 2, 4, 6, 8
13	0	all	1, 3, 5, 7, 9	45	2	all	1, 3, 5, 7, 9
14	0	all	All	46	N.A.	N.A.	N.A.
15	0	even	9	47	2	even	9
16	1	even	1	48	3	even	1
17	1	even	4	49	3	even	4
18	1	even	7	50	3	even	7
19	1	all	1	51	3	all	1
20	1	all	4	52	3	all	4
21	1	all	7	53	3	all	7
22	1	all	1, 6	54	3	all	1, 6
23	1	all	2 ,7	55	3	all	2 ,7
24	1	all	3, 8	56	3	all	3, 8
25	1	all	1, 4, 7	57	3	all	1, 4, 7
26	1	all	2, 5, 8	58	3	all	2, 5, 8
27	1	all	3, 6, 9	59	3	all	3, 6, 9
28	1	all	0, 2, 4, 6, 8	60	N.A.	N.A.	N.A.
29	1	all	1, 3, 5, 7, 9	61	N.A.	N.A.	N.A.
30	N.A.	N.A.	N.A.	62	N.A.	N.A.	N.A.
31	1	even	9	63	3	even	9

N.A.: not applicable

Table 9.2. *Configuration of the subframe allocated to the PRACH physical channel: type-1 time frame*

The parameter $t_{RA}^{(2)}$ shows the number of the subframe used by the PRACH physical channel.

The numbering starts at each commutation point from the downlink to the uplink direction.

The parameter value $t_{\mathrm{RA}}^{(2)}$ depends on the configuration index of the PRACH physical channel.

The position $n_{\mathrm{PRB}}^{\mathrm{RA}}$ of the physical channel is provided by the following formulas:

For the formats 0 to 3:

$$n_{\mathrm{PRB}}^{\mathrm{RA}} = n_{\mathrm{PRB\ offset}}^{\mathrm{RA}} + 6\left\lfloor \frac{f_{\mathrm{RA}}}{2} \right\rfloor, \text{ if } f_{\mathrm{RA}} \bmod 2 = 0$$

$$n_{\mathrm{PRB}}^{\mathrm{RA}} = N_{\mathrm{RB}}^{\mathrm{UL}} - 6 - n_{\mathrm{PRB\ offset}}^{\mathrm{RA}} - 6\left\lfloor \frac{f_{\mathrm{RA}}}{2} \right\rfloor, \text{ otherwise}$$

f_{RA} corresponds to a frequency resource index whose value depends on the configuration index of the PRACH physical channel;

$N_{\mathrm{RB}}^{\mathrm{UL}}$ corresponds to the bandwidth of the radio channel for the uplink direction, expressed in the number of resource blocks allocated.

For format 4:

$$n_{\mathrm{PRB}}^{\mathrm{RA}} = 6 f_{\mathrm{RA}}, \text{ if } \left((n_{\mathrm{f}} \bmod 2) \times (2 - N_{\mathrm{SP}}) + t_{\mathrm{RA}}^{(1)} \right) \bmod 2 = 0$$

$$n_{\mathrm{PRB}}^{\mathrm{RA}} = N_{\mathrm{RB}}^{\mathrm{UL}} - 6(f_{\mathrm{RA}} + 1), \text{ otherwise}$$

n_{f} corresponds to the number of the time frame;

N_{SP} corresponds to the number of commutation points from the downlink to the uplink direction;

$t_{\mathrm{RA}}^{(1)}$ corresponds to a parameter which shows if the PRACH channel is in the first half of the frame (value at ZERO) or the second (value at ONE).

The parameter value $t_{\mathrm{RA}}^{(1)}$ depends on the configuration index of the PRACH physical channel.

Index	Format Preamble	Density	Version	Index	Format Preamble	Density	Version
0	0	0.5	0	32	2	0.5	2
1	0	0.5	1	33	2	1	0
2	0	0.5	2	34	2	1	1
3	0	1	0	35	2	2	0
4	0	1	1	36	2	3	0
5	0	1	2	37	2	4	0
6	0	2	0	38	2	5	0
7	0	2	1	39	2	6	0
8	0	2	2	40	3	0.5	0
9	0	3	0	41	3	0.5	1
10	0	3	1	42	3	0.5	2
11	0	3	2	43	3	1	0
12	0	4	0	44	3	1	1
13	0	4	1	45	3	2	0
14	0	4	2	46	3	3	0
15	0	5	0	47	3	4	0
16	0	5	1	48	4	0.5	0
17	0	5	2	49	4	0.5	1
18	0	6	0	50	4	0.5	2
19	0	6	1	51	4	1	0
20	1	0.5	0	52	4	1	1
21	1	0.5	1	53	4	2	0
22	1	0.5	2	54	4	3	0
23	1	1	0	55	4	4	0
24	1	1	1	56	4	5	0
25	1	2	0	57	4	6	0
26	1	3	0	58	N.A.	N.A.	N.A.
27	1	4	0	59	N.A.	N.A.	N.A.
28	1	5	0	60	N.A.	N.A.	N.A.
29	1	6	0	61	N.A.	N.A.	N.A.
30	2	0.5	0	62	N.A.	N.A.	N.A.
31	2	0.5	1	63	N.A.	N.A.	N.A.

N.A.: not applicable

Table 9.3. *Density of the subframe allocated to the PRACH physical channel: type-2 time frame*

9.1.4. *Sequence generation of the preamble*

The preamble is produced from a Zadoff-Chu sequence that corresponds to the following formula:

$$x_u(n) = e^{-j\frac{\pi un(n+1)}{N_{ZC}}}, \quad 0 \le n \le N_{ZC} - 1$$

N_{ZC} corresponds to the length of the Zadoff-Chu sequence;

u corresponds to the number of the root.

Starting from the Zadoff-Chu sequence, 64 orthogonal sequences are produced by a cyclic shift C_v: $x_{u,v}(n) = x_u((n + C_v) \bmod N_{ZC})$.

The 64 preambles are divided into two groups:

– the group for which the random access procedure is carried out with contention issues;

– the group for which the random access procedure is carried out without contention issues.

The group involved in the procedure with contention issues is divided into two subgroups:

– the subgroup for which the quantity of data to be transmitted by the mobile is low or when the mobile cover is of poor quality;

– the subgroup for which the quantity of data to be transmitted by the mobile is high and when the mobile cover is of good quality.

9.1.5. *Power control*

The power used for the transmission of the PRACH physical channel is defined by the following formula:

$$P_{PRACH} = \min\{ P_{CMAX}, \text{PREAMBLE\_RECEIVED\_TARGET\_POWER} + PL \}$$

P_{CMAX} corresponds to the configured maximum transmit power of the mobile.

PREAMBLE_RECEIVED_TARGET_POWER corresponds to the power of the PRACH physical channel received at the eNB entity level.

PREAMBLE_RECEIVED_TARGET_POWER=PREAMBLE_INITIAL_RECEIVED_TARGET_POWER+DELTA_PREAMBLE+(PREAMBLE_TRANSMISSION_COUNTER-1)**powerRampingStep*

The value of the parameter PREAMBLE_INITIAL_RECEIVED_TARGET_POWER is shown by the SIB2 system information or the RRC *ConnectionReconfiguration* message.

The value of the parameter DELTA_PREAMBLE depends on the preamble format.

The value of the parameter PREAMBLE_TRANSMISSION_COUNTER is initialized at ONE at the start of the random access procedure, then increased by a unit in case there is no response from the eNB entity.

The value of the parameter *powerRampingStep* is shown by the SIB2 system information or the RRC *ConnectionReconfiguration* message. This parameter determines the increase of transmission power of the PRACH physical channel in case of non-response of the eNB entity.

PL corresponds to an assessment of the propagation loss calculated by the mobile.

9.2. PUCCH physical channel

The Physical Uplink Control Channel (PUCCH) uses three types of format for transporting the Uplink Control Information (UCI):

– formats 1, 1a and 1b transport the UCI information in relation to the scheduling request to get resources on the Physical Uplink Shared Channel (PUSCH) and to the positive ACK or negative NACK acknowledgment of the received data on the Physical Uplink Shared Channel (PDSCH), corresponding to the Hybrid Automatic Repeat reQuest (HARQ) mechanism;

– formats 2, 2a and 2b transport the UCI information in relation to the status of the signal received on the PDSCH physical channel and to the positive ACK or negative NACK acknowledgment;

– format 3 transports the same information that format 1 by adapting to the aggregation of Carrier Component (CC) introduced in Release 10.

For Release 8 and 9, the PUCCH physical channel transports the UCI information only if the mobile does not transmit on the PUSCH physical channel.

If the mobile transmits on the PUSCH physical channel, the UCI information is multiplexed with the Uplink Shared Channel (UL-SCH).

Release 10 introduced the simultaneous transmission of the PUCCH and PUSCH physical channels.

9.2.1. *UCI information*

The Scheduling Request (SR) information involves the request in order to get resources for the PUSCH physical channel.

The HARQ Indicator (HI) information involves the positive ACK or negative NACK acknowledgment of data received on the PDSCH physical channel.

The Channel Status Information (CSI) regroups the information related to the status report of the signal received on the PDSCH physical channel:

– the Channel Quality Indicator (CQI) represents the Modulation and Coding Scheme (MCS) recommended for the PDSCH physical channel;

– the Rank Indicator (RI) determines the number of spatial layers recommended for the PDSCH physical channel;

– the Precoder Matrix Indicator (PMI) provides indications related to the precoding matrix in the case of using spatial multiplexing, in closed loop.

9.2.1.1. *UCI information formats*

Format 1 takes into account the SR information for which no bit is transmitted, the evolved NodeB (eNB) entity only detects the presence of energy in the PUCCH channel.

Format 1a takes into account the HI information corresponding to one bit of positive ACK or negative NACK acknowledgment of a transport block received on the PDSCH physical channel.

Format 1a can couple the HI and SR information, only for the Frequency Division Duplex (FDD) mode.

Format 1b takes into account, for Release 8 and 9, the HI information corresponds to two bits of positive ACK or negative NACK acknowledgment of two transport blocks received on the PDSCH physical channel.

Format 1b can couple the HI and SR information.

Format 1b takes into account, for Release 10, the HI information corresponding to four bits of positive ACK or negative NACK acknowledgment of two transport blocks received on the PDSCH physical channel, for two aggregated radio channels.

Format 1b limits the number of bits in 2, used by the selection of the PUCCH physical channel.

For the Time Division Duplex (TDD) mode only, format 1b takes into account the HI information corresponding to four bits of positive ACK or negative NACK acknowledgment.

Format 2 takes into account CSI information corresponding to 20 bits related to the status of the PDSCH physical channel.

Format 2a (2b respectively) can couple CSI and HI information coded on 1 bit (respectively coded on 2 bits), for the FDD mode only.

Format 3 takes into account the HI information corresponding to:

– ten bits of positive ACK or negative NACK acknowledgment of two transport blocks received on the PDSCH physical channel, for five aggregated radio channels, for the FDD mode;

– twenty bits of positive ACK or negative NACK acknowledgment of two transport blocks received on the PDSCH physical channel, for five aggregated radio channels, for the TDD mode.

Table 9.4 recapitulates the mapping of the formats and UCI information.

<table>
<tr><th>PUCCH format</th><th>FDD/TDD</th><th>Normal cyclic prefix</th><th>Extended cyclic prefix</th></tr>
<tr><td>1</td><td>FDD/TDD</td><td colspan="2">SR</td></tr>
<tr><td rowspan="2">1a</td><td>FDD/TDD</td><td colspan="2">$1 \times$ HI</td></tr>
<tr><td>FDD</td><td colspan="2">$1 \times$ HI + SR</td></tr>
<tr><td rowspan="2">1b</td><td>FDD/TDD</td><td colspan="2">$2 \times$ HI, or
$2 \times$ HI + SR, or
$4 \times$ HI for 2CC, with selection of PUCCH channels</td></tr>
<tr><td>TDD</td><td colspan="2">$4 \times$ HI with selection of PUCCH channels</td></tr>
<tr><td>2</td><td>FDD/TDD</td><td>CSI</td><td>CSI, or
CSI + $2 \times$ HI</td></tr>
<tr><td>2a</td><td>FDD/TDD</td><td>CSI + $1 \times$ HI</td><td rowspan="2">Not applicable</td></tr>
<tr><td>2b</td><td>FDD/TDD</td><td>CSI + $2 \times$ HI</td></tr>
<tr><td rowspan="2">3</td><td>FDD</td><td colspan="2">$10 \times$ HI, or
$10 \times$ HI + SR</td></tr>
<tr><td>TDD</td><td colspan="2">$20 \times$ HI, or
$20 \times$ HI + SR</td></tr>
</table>

Table 9.4. *Mapping of the formats and the UCI information.*

9.2.1.2. *CSI reports*

The report of the CSI information can be periodic or aperiodic.

The aperiodic report is always transferred in the PUSCH physical channel.

The periodic report is transferred in the PUCCH physical channel in the two following cases:

– no resource is allocated to the mobile in the PUSCH physical channel;

– a resource is allocated to the mobile and the simultaneous transmission of the PUCCH and PUSCH physical channels is possible (Release 10).

Otherwise, the periodic report is transmitted in the PUSCH physical channel.

The aperiodic report allows for feedback related to the CSI information in relation to the totality of the bandwidth of the radio channel or to a part of the bandwidth of the radio channel defined by the mobile or the eNB entity.

The transfer of the aperiodic report is triggered by the following messages:

– the Downlink Control Information (DCI) in the 0 or 4 format transmitted in the Physical Downlink Control Channel (PDCCH);

– the Random Access Response (RAR) message transmitted in the PDSCH physical channel.

The periodic report allows for feedback to the CSI information in relation to the totality of the bandwidth of the radio channel or to a part of the bandwidth of the radio channel defined only by the eNB entity.

The type of transfer of the aperiodic or periodic report is communicated to the mobile by the RRC messages *ConnectionSetup, ConnectionReconfiguration* or *ConnectionReconfiguration.*

The modes of transfer of the reports are constructed depending on the type of return of the CQI and PMI information (Tables 9.5 and 9.6).

		PMI return		
		No PMI	Single PMI	Multiple PMI
Return CQI	Wide band			Mode 1-2
	Subband by UE	Mode 2-0		Mode 2-2
	Subband by ENB	Mode 3-0	Mode 3-1	

Table 9.5. *Transfer modes of the aperiodic reports*

		PMI return	
		No PMI	Single PMI
Return CQI	Wide band	Mode 1-0	Mode 1-1
	Subband by ENB	Mode 2-0	Mode 2-1

Table 9.6. *Transfer modes of the periodic reports*

9.2.2. *Formats 1, 1a and 1b*

The processing associated with the PUCCH physical channel for formats 1, 1a and 1b is summarized in Figure 9.3.

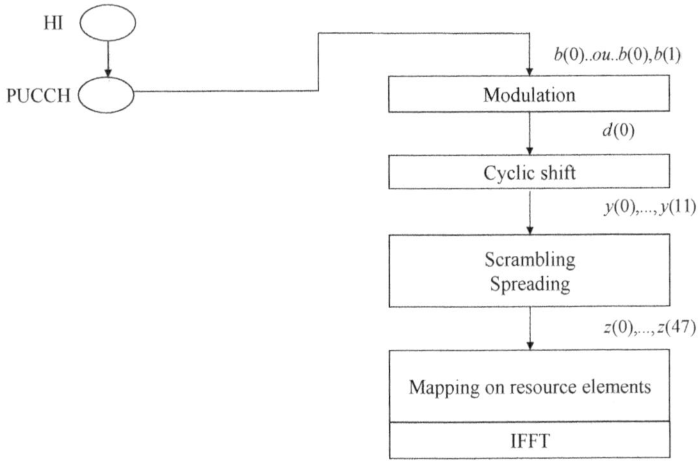

Figure 9.3. *Processing associated with the PUCCH physical channel: formats 1, 1a and 1b*

9.2.2.1. *Modulation*

For format 1a, the transmission of the HI information bit uses the Binary Phase-Shift Keying (BPSK) modulation to produce a symbol $d(0)$.

For format 1b, the transmission of the 2 HI information bits uses the Quadrature Phase-Shift Keying (QPSK) modulation to produce a symbol $d(0)$.

For format 1, the value of $d(0)$ is equal to 1.

9.2.2.2. *Cyclic shift*

The symbol $d(0)$ is multiplied by a sequence of cyclic shift $r_{u,v}^{(\alpha)}(n)$ for generating the sequence $y(n)$:

$$y(n) = d(0) \cdot r_{u,v}^{(\alpha)}(n), \qquad n = 0,1,...,N_{\text{seq}}^{\text{PUCCH}} - 1$$

$N_{\text{seq}}^{\text{PUCCH}}$ corresponds to the number of the cyclic shift sequence: $N_{\text{seq}}^{\text{PUCCH}} = 12$

The base sequence is generated from the following formula:

$$r_{\text{PUSCH}} = r_{u,v}^{(\alpha)}(n) = e^{j\alpha n} \cdot \overline{r}_{u,v}(n)$$

– α corresponds to the cyclic shift applied to the base sequence $\overline{r}_{u,v}(n)$;

– α can take 12 values of cyclic shift;

– u corresponds to the group number of the base sequence;

– u can take 30 values, from 0 to 29;

– v corresponds to the index of the base sequence in the group;

– v otherwise it takes a single value, 0;

– n takes the value between 0 and $N_{\text{seq}}^{\text{PUCCH}} - 1$, $N_{\text{seq}}^{\text{PUCCH}}$ is the length of the PUCCH physical signal.

The base sequence $\overline{r}_{u,v}(n)$ is generated from the formula:

$$\overline{r}_{u,v}(n) = e^{j\phi(n)\pi/4}, \quad 0 \le n \le N_{\text{seq}}^{\text{PUCCH}} - 1$$

The value of $\phi(n)$ depends on the group number of the base sequence u.

The group number of the base sequence is determined from the following formula:

$$u = \left(f_{\text{gh}}(n_{\text{s}}) + f_{\text{ss}}\right) \bmod 30 .$$

$f_{\text{gh}}(n_{\text{s}})$ is a random number between 0 and 29 if the group hopping is activated. Otherwise, it takes the value equal to 0.

$f_{\text{gh}}(n_{\text{s}})$ depends on the value of the time slot n_{s} and of the value $N_{\text{ID}}^{\text{cell}}$ of the Physical-layer Cell Identity (PCI).

$f_{ss} = \left(N_{ID}^{cell} + \Delta_{ss}\right) \bmod 30$, the $\Delta_{ss} \in \{0,1,...,29\}$ value is shown in the message of System Information Block 2 (SIB2).

9.2.2.3. *Scrambling and spreading*

The sequence $y(0),...,y(N_{seq}^{PUCCH}-1)$ is then multiplied by the scrambling $S(n_s)$ and spreading $w(m)$ sequences to obtain the following sequence:

$$z\left(m' \cdot N_{SF}^{PUCCH} \cdot N_{seq}^{PUCCH} + m \cdot N_{seq}^{PUCCH} + n\right) = S(n_s) \cdot w(m) \cdot y(n)$$

$S(n_s)$ is equal to $S(0)=1$ for the even time slots and to $S(1)=e^{j\pi/2}$ for the odd time slots.

$w(m)$ is an orthogonal sequence, with $m = 0,..., N_{SF}^{PUCCH}-1$.

N_{SF}^{PUCCH} corresponds to the number of orthogonal sequence bits: $N_{SF}^{PUCCH}=4$ m' takes two values, 0 for the first Resource Block (RB) and 1 for the second resource block.

9.2.2.4. *Mapping on the resource elements*

The mapping on the Resource Elements (RE) of the two resource blocks, for a normal cyclic prefix, is described in Figure 9.4.

For an extended cyclic prefix, the number of symbols allocated to the PUCCH physical channel is not modified.

The cyclic shift sequence $r_{u,v}^{(\alpha)}(n)$ and the orthogonal sequence $w(m)$ are specific to each mobile and they are used for multiplexing several mobiles in the two resource blocks.

The number of mobiles using the same resource blocks depends on the number of cyclic shift sequences, the number of orthogonal sequences is fixed at 3:

– 12 cyclic shift sequences allow the multiplexing of 36 mobiles;

– 6 cyclic shift sequences allow the multiplexing of 18 mobiles;

– 4 cyclic shift sequences allow the multiplexing of 12 mobiles.

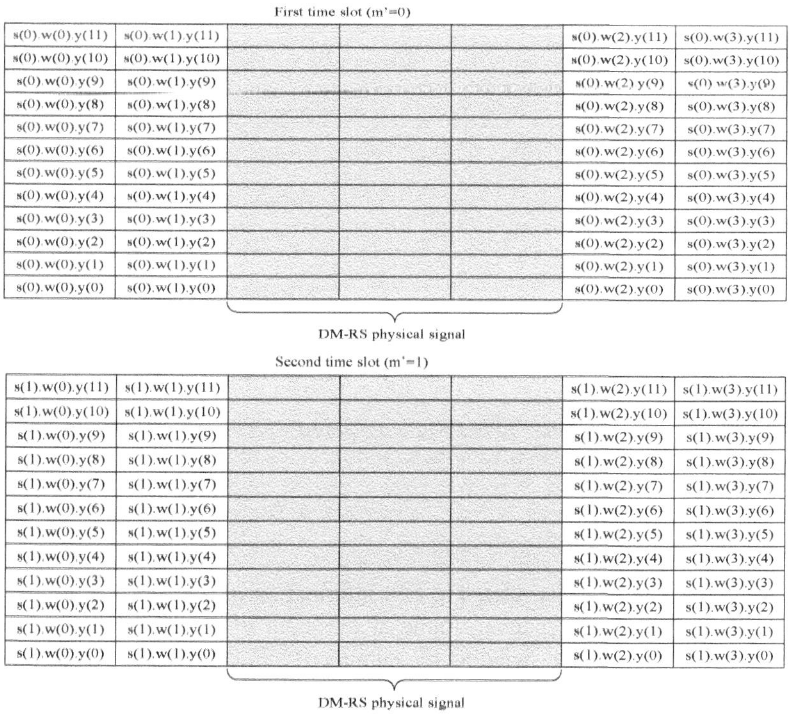

Figure 9.4. *Mapping of the PUCCH physical channel: formats 1, 1a and 1b*

The transmission of the Sounding Reference Signal (SRS) reference signal can pre-empt the last symbol of the subframe, and in this case, an orthogonal sequence of three bits is applied, which does not modify the number of multiplexed mobiles.

9.2.3. *Formats 2, 2a and 2b*

The processing associated with the PUCCH physical channel, for formats 2, 2a and 2b, is summarized in Figure 9.5.

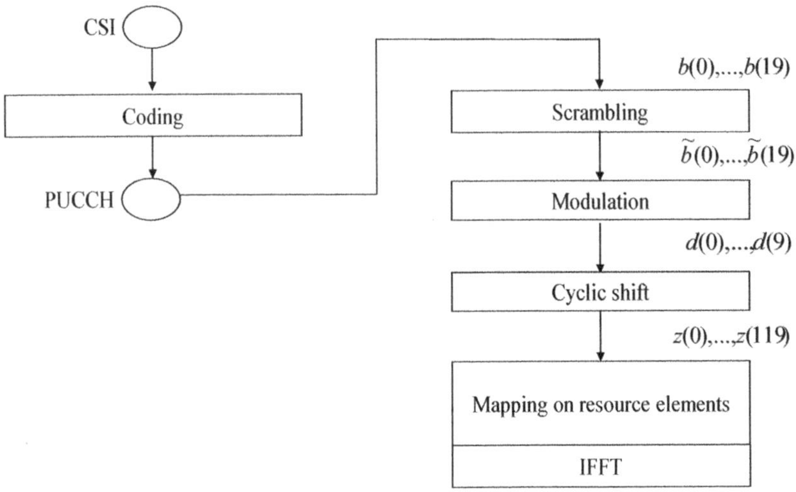

Figure 9.5. *Processing associated with the PUCCH physical channel: formats 2, 2a and 2b*

9.2.3.1. *Coding of CSI information*

The information bits related to CSI information of the signal received on the PDSCH physical channel, found at the input of the encoder are called $a_0, a_1, a_2, a_3, ..., a_{A-1}$ A designating the number of bits to be coded.

The bits found at the output of the encoder are called $b_0, b_1, b_2, b_3, ..., b_{B-1}$, $b_i = \sum_{n=0}^{A-1} (a_n \cdot M_{i,n}) \bmod 2$, with the index i = 0, 1, 2,..., B-1 with B = 20 and sequences $M_{i,n}$ defined in Table 9.7.

For formats 2a and 3b, in the case of normal cyclic prefix, the multiplexing of the CSI and HI information occurs after the coding of the CSI information. We obtain 21 bits for format 2a or 22 bits for format 2b.

For format 2, in the case of extended cyclic prefix, the multiplexing of the CSI and HI information occurs before the coding of the CSI information. We therefore obtain 20 bits.

i	$M_{i,0}$	$M_{i,1}$	$M_{i,2}$	$M_{i,3}$	$M_{i,4}$	$M_{i,5}$	$M_{i,6}$	$M_{i,7}$	$M_{i,8}$	$M_{i,9}$	$M_{i,10}$	$M_{i,11}$	$M_{i,12}$
0	1	1	0	0	0	0	0	0	0	0	1	1	0
1	1	1	1	0	0	0	0	0	0	1	1	1	0
2	1	0	0	1	0	0	1	0	1	1	1	1	1
3	1	0	1	1	0	0	0	0	1	0	1	1	1
4	1	1	1	1	0	0	0	1	0	0	1	1	1
5	1	1	0	0	1	0	1	1	1	0	1	1	1
6	1	0	1	0	1	0	1	0	1	1	1	1	1
7	1	0	0	1	1	0	0	1	1	0	1	1	1
8	1	1	0	1	1	0	0	1	0	1	1	1	1
9	1	0	1	1	1	0	1	0	0	1	1	1	1
10	1	0	1	0	0	1	1	1	0	1	1	1	1
11	1	1	1	0	0	1	1	0	1	0	1	1	1
12	1	0	0	1	0	1	0	1	1	1	1	1	1
13	1	1	0	1	0	1	0	1	0	1	1	1	1
14	1	0	0	0	1	1	0	1	0	0	1	0	1
15	1	1	0	0	1	1	1	1	0	1	1	0	1
16	1	1	1	0	1	1	1	0	0	1	0	1	1
17	1	0	0	1	1	1	0	0	1	0	0	1	1
18	1	1	0	1	1	1	1	1	0	0	0	0	0
19	1	0	0	0	0	1	1	0	0	0	0	0	0

Table 9.7. *Sequences Mi, n*

9.2.3.2. *Scrambling*

The sequence of bits $b(0),...,b(M_{\text{bit}}-1)$, $M_{\text{bit}}=20$ designating the number of bits, is scrambled by the sequence $c\ (i)$.

It results in the bit sequence $\tilde{b}(0),...,\tilde{b}(M_{\text{bit}}-1)$, with $\tilde{b}(i)=\left(b(i)+c(i)\right)\text{mod}\,2$.

The scrambling sequence is made of a pseudo-random sequence that uses a 31-bit polynomial.

The initial value of the scrambling sequence c_{init} is calculated from the following formula:

$$c_{\text{init}}=\left(\lfloor n_{\text{s}}/2\rfloor+1\right)\cdot\left(2N_{\text{ID}}^{\text{cell}}+1\right)\cdot 2^{16}+n_{\text{RNTI}}$$

n_{RNTI} corresponds to the Radio Network Temporary Identifier (RNTI) allocated to the mobile during connection.

n_{s} corresponds to the number of the time slot of the time frame.

$N_{\text{ID}}^{\text{cell}}$ corresponds to the Physical-layer Cell Identity (PCI).

9.2.3.3. *Modulation*

For formats 2, 2a and 2b, the sequence transmission $\tilde{b}(0),...,\tilde{b}(M_{\text{bit}}-1)$ uses the QPSK modulation to produce the sequence $d(0),...,d(M_{\text{symb}}-1)$, $M_{\text{symb}}=10$.

For format 2a, the transmission of the HI information bit uses the BPSK modulation to produce an eleventh symbol transmitted in the Demodulation Reference Signal (DM-RS) reference signal.

For format 2b, the transmission of the HI information bit uses the QPSK modulation to produce an eleventh symbol transmitted in the DM-RS reference signal.

9.2.3.4. *Cyclic shift*

Each symbol of the sequence $d(0),...,d(9)$ is multiplied by a sequence of cyclic shift $r_{u,v}^{(\alpha)}(n)$ to generate the sequence $z(0),...,z(119)$:

$$z(N_{\text{seq}}^{\text{PUCCH}}\cdot n+i)=d(n)\cdot r_{u,v}^{(\alpha)}(i),\text{ with } n=0,1,...,9 \text{ and } i=0,1,...,N_{\text{sc}}^{\text{RB}}-1$$

N_{sc}^{RB} corresponds to the number of subcarriers of a resource block in the frequency domain: $N_{sc}^{RB} = 12$.

9.2.3.5. *Mapping on the resource components*

The mapping on the resource elements of the two resource blocks is described in Figure 9.6.

First time slot

	DM-RS physical signal				DM-RS physical signal	
d(0).r(11)		d(1).r(11)	d(2).r(11)	d(3).r(11)		d(4).r(11)
d(0).r(10)		d(1).r(10)	d(2).r(10)	d(3).r(10)		d(4).r(10)
d(0).r(9)		d(1).r(9)	d(2).r(9)	d(3).r(9)		d(4).r(9)
d(0).r(8)		d(1).r(8)	d(2).r(8)	d(3).r(8)		d(4).r(8)
d(0).r(7)		d(1).r(7)	d(2).r(7)	d(3).r(7)		d(4).r(7)
d(0).r(6)		d(1).r(6)	d(2).r(6)	d(3).r(6)		d(4).r(6)
d(0).r(5)		d(1).r(5)	d(2).r(5)	d(3).r(5)		d(4).r(5)
d(0).r(4)		d(1).r(4)	d(2).r(4)	d(3).r(4)		d(4).r(4)
d(0).r(3)		d(1).r(3)	d(2).r(3)	d(3).r(3)		d(4).r(3)
d(0).r(2)		d(1).r(2)	d(2).r(2)	d(3).r(2)		d(4).r(2)
d(0).r(1)		d(1).r(1)	d(2).r(1)	d(3).r(1)		d(4).r(1)
d(0).r(0)		d(1).r(0)	d(2).r(0)	d(3).r(0)		d(4).r(0)

Second time slot

	DM-RS physical signal				DM-RS physical signal	
d(5).r(11)		d(6).r(11)	d(7).r(11)	d(8).r(11)		d(9).r(11)
d(5).r(10)		d(6).r(10)	d(7).r(10)	d(8).r(10)		d(9).r(10)
d(5).r(9)		d(6).r(9)	d(7).r(9)	d(8).r(9)		d(9).r(9)
d(5).r(8)		d(6).r(8)	d(7).r(8)	d(8).r(8)		d(9).r(8)
d(5).r(7)		d(6).r(7)	d(7).r(7)	d(8).r(7)		d(9).r(7)
d(5).r(6)		d(6).r(6)	d(7).r(6)	d(8).r(6)		d(9).r(6)
d(5).r(5)		d(6).r(5)	d(7).r(5)	d(8).r(5)		d(9).r(5)
d(5).r(4)		d(6).r(4)	d(7).r(4)	d(8).r(4)		d(9).r(4)
d(5).r(3)		d(6).r(3)	d(7).r(3)	d(8).r(3)		d(9).r(3)
d(5).r(2)		d(6).r(2)	d(7).r(2)	d(8).r(2)		d(9).r(2)
d(5).r(1)		d(6).r(1)	d(7).r(1)	d(8).r(1)		d(9).r(1)
d(5).r(0)		d(6).r(0)	d(7).r(0)	d(8).r(0)		d(9).r(0)

Figure 9.6. *Mapping of the PUCCH physical channel: formats 2, 2a and 2b*

The cyclic shift sequence $r_{u,v}^{(\alpha)}(n)$ is specific to each mobile and it is used for multiplexing 12 mobiles in the two resource blocks.

The SRS reference signal cannot pre-empt the last symbol of the subframe allocated in the PUCCH channel for formats 2, 2a and 2b.

9.2.4. *Format 3*

The processing associated with the PUCCH physical channel, for the format 3, is summarized in Figure 9.7.

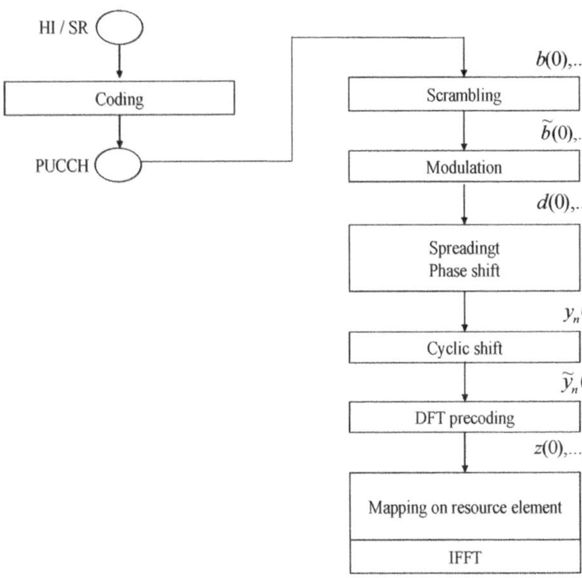

Figure 9.7. *Processing associated with the PUCCH physical channel: format 3*

9.2.4.1. *Coding of the HI and SR information*

For format 3, the HI and SR information can be multiplexed to make 11 bits for the FDD mode and 21 bits for the TDD mode; these bits are coded to generate 48 bits.

9.2.4.2. *Scrambling*

The sequence of bits $b(0),...,b(M_{\text{bit}}-1)$, $M_{\text{bit}}=48$ designating the number of bits, is scrambled by the sequence $c\ (i)$.

It results in the bit sequence $\tilde{b}(0),...,\tilde{b}(M_{\text{bit}}-1)$, with $\tilde{b}(i)=\left(b(i)+c(i)\right)\bmod 2$.

The scrambling sequence is made of a pseudo-random sequence that uses a 31-bit polynomial.

The initial value of the scrambling sequence c_{init} is calculated from the following formula:

$$c_{\text{init}}=\left(\lfloor n_s/2\rfloor+1\right)\cdot\left(2N_{\text{ID}}^{\text{cell}}+1\right)\cdot 2^{16}+n_{\text{RNTI}}$$

9.2.4.3. *Modulation*

For format 3, the sequence transmission $\tilde{b}(0),...,\tilde{b}(M_{\text{bit}}-1)$ uses the QPSK modulation to produce the sequence $d(0),...,d(M_{\text{symb}}-1)$, $M_{\text{symb}}=24$.

9.2.4.4. *Spreading and phase shift*

The symbol sequence $d(0),...,d23$ is divided into two parts, each part is allocated to a time slot.

Each sequence of 12 symbols $d(0),...,d11$ or $d(12),...,d23$ is repeated 5 times, and a phase shift $e^{j\pi\lfloor n_{\text{cs}}^{\text{cell}}(n_s,l)/64\rfloor/2}$ is applied to each replica.

$n_{\text{cs}}^{\text{cell}}(n_s,l)$ is a generated value from a pseudo-random sequence and depending on the physical identity of the cell $N_{\text{ID}}^{\text{cell}}$, the time slot number n_s and the symbol number l.

The 5 replicas of each time slot are multiplied by an orthogonal 5 bit code, specific to the time slot, $w_0(\overline{n})$ for the first time slot and $w_1(\overline{n})$ for the second time slot.

$$\overline{n}=n\bmod 5$$

We obtain the sequence $y_n(i)$, with i variant from 0 to 11 for the first time slot and from 12 to 23 for the second time slot, and n variant from 0 to 9.

$$y_n(i) = w_0(\overline{n}) \cdot e^{j\pi\left\lfloor n_{cs}^{cell}(n_s,l)/64\right\rfloor/2} \cdot d(i) \text{ for the first time slot;}$$

$$y_n(i) = w_1(\overline{n}) \cdot e^{j\pi\left\lfloor n_{cs}^{cell}(n_s,l)/64\right\rfloor/2} \cdot d(i+12) \text{ for the second time slot.}$$

9.2.4.5. *Cyclic shift*

A cyclic shift is applied to the sequence $y_n(i)$ to generate the sequence $\tilde{y}_n(i)$ with $\tilde{y}_n(i) = y_n\left(\left(i + n_{cs}^{cell}(n_s,l)\right) \bmod N_{sc}^{RB}\right)$.

N_{sc}^{RB} corresponds to the number of subcarriers per resource block, in the frequency domain.

9.2.4.6. *Precoding*

A Discrete Fourier Transform (DFT) is applied to the sequence $\tilde{y}_n(i)$ for generating the sequence $z(0),...,z(119)$:

$$z(n \cdot N_{sc}^{RB} + k) = \frac{1}{\sqrt{N_{sc}^{RB}}} \sum_{i=0}^{N_{sc}^{RB}-1} \tilde{y}_n(i) e^{-j\frac{2\pi ik}{N_{sc}^{RB}}}$$

$$n = 0,...,9 \text{ and } k = 0,..., N_{sc}^{RB} - 1$$

9.2.4.7. *Mapping on the resource elements*

The mapping on the resource elements of the two resource blocks is described in Figure 9.8.

The orthogonal sequences $w_0(\overline{n})$ and $w_1(\overline{n})$ are specific to each mobile and are used for multiplexing 5 mobiles in two resource blocks.

The transmission of the SRS reference signal can pre-empt the last symbol of the subframe, and in this case, an orthogonal sequence of 4 bits is specified, which reduces the number of multiplexed mobiles in the two resource blocks.

n=0		n=1	n=2	n=3		n=4
z(11)		z(23)	z(35)	z(47)		z(59)
z(10)		z(22)	z(34)	z(46)		z(58)
z(9)		z(21)	z(33)	z(45)		z(57)
z(8)		z(20)	z(32)	z(44)		z(56)
z(7)		z(19)	z(31)	z(43)		z(55)
z(6)		z(18)	z(30)	z(42)		z(54)
z(5)		z(17)	z(29)	z(41)		z(53)
z(4)		z(16)	z(28)	z(40)		z(52)
z(3)		z(15)	z(27)	z(39)		z(51)
z(2)		z(14)	z(26)	z(38)		z(50)
z(1)		z(13)	z(25)	z(37)		z(49)
z(0)		z(12)	z(24)	z(36)		z(48)

DM-RS physical signal — First time slot — DM-RS physical signal

n=5		n=6	n=7	n=8		n=9
z(71)		z(83)	z(95)	z(107)		z(119)
z(70)		z(82)	z(94)	z(106)		z(118)
z(69)		z(81)	z(93)	z(105)		z(117)
z(68)		z(80)	z(92)	z(104)		z(116)
z(67)		z(79)	z(91)	z(103)		z(115)
z(66)		z(78)	z(90)	z(102)		z(114)
z(65)		z(77)	z(89)	z(101)		z(113)
z(64)		z(76)	z(88)	z(100)		z(112)
z(63)		z(75)	z(87)	z(99)		z(11)
z(62)		z(74)	z(86)	z(98)		z(110)
z(61)		z(73)	z(85)	z(97)		z(109)
z(60)		z(72)	z(84)	z(96)		z(108)

DM-RS physical signal — Second time slot — DM-RS physical signal

Figure 9.8. *Mapping of the PUCCH physical channel: format 3*

9.2.5. Configuration of the resource blocks

The two resource blocks allocated to the PUCCH physical channel are located:

– in the frequency domain, at the two extremities of the radio channel;

– in the time domain, in two time slots of a subframe.

The first time slot (respectively the second time slot) is located in the lowest frequency band (respectively the highest frequency band) of the radio channel (Figure 9.9).

The first PUCCH physical channels are allocated to formats 2, 2a and 2b (Figure 9.9).

The next PUCCH physical channels are allocated to formats 1, 1a and 2b (Figure 9.9).

The PUCCH physical channel shared between type-1 and type-2 formats is inserted between the PUCCH physical channels allocated, on the one hand, to type 2, and, on the other hand, to type 1 (Figure 9.9).

The PUCCH physical channels in format 3 are located after the PUCCH physical channels in formats 2, 2a, 2b (Figure 9.9).

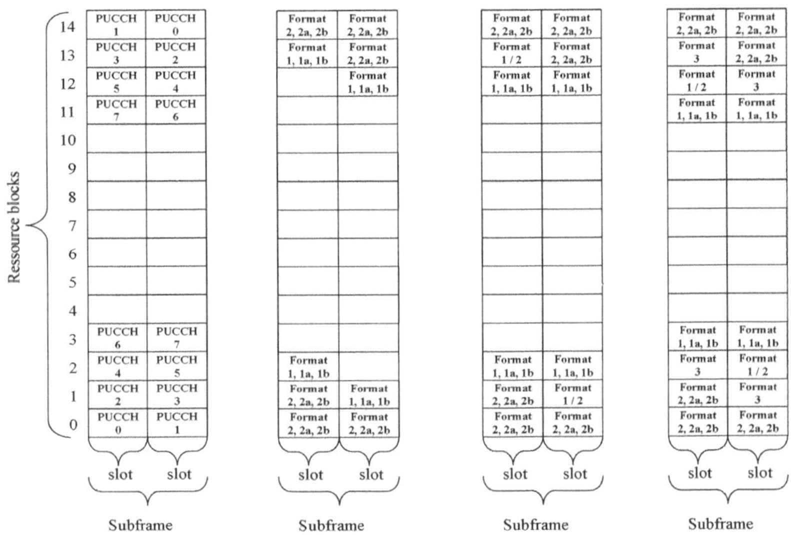

Figure 9.9. *Configuration of the resource blocks allocated to the PUCCH physical channel*

The allocated resource blocks in the different formats depend on two parameters:

– $N_{\mathrm{RB}}^{(2)}$ is the number of resource blocks allocated to formats 2, 2a and 2b in each time slot;

– $N_{cs}^{(1)}$ is the number of sequences of cyclic shift allocated to formats 1, 1a and 1b, when the two resource blocks are shared by the type-1 and type-2 formats.

The parameter $N_{cs}^{(1)}$ is a multiple of the parameter Δ_{shift}^{PUCCH}, which can take a value in the interval{0, 1, …,7}.

The parameters $N_{RB}^{(2)}$, $N_{cs}^{(1)}$ and Δ_{shift}^{PUCCH} are communicated by the SIB2 information system.

9.2.6. *Allocation of the PUCCH physical channel*

The PUCCH physical channel in formats 1, 1a or 1b allocated to the mobile is located by the parameter $n_{PUCCH}^{(1)}$, calculated from the following formula:

$$n_{PUCCH}^{(1)} = n_{CCE} + N_{PUCCH}^{(1)}$$

n_{CCE} corresponds to the number of the first Channel Control Element (CCE) of the PDCCH physical channel, in which the resource allocated to the mobile in the PDSCH physical channel is described.

$N_{PUCCH}^{(1)}$ corresponds to a parameter communicated in the SIB2 system information.

To transmit the information of the SR scheduling request, the mobile uses the parameter $n_{PUCCH}^{(1)}$ communicated to the mobile by the RRC *ConnectionReconfiguration* message.

In the case of format 1b, for which 4 HI information must be transferred, the mobile locates a PUCCH physical channel among four values $n_{PUCCH}^{(1)}$, each channel represents a combination of positive and negative acknowledgments.

The PUCCH physical channel in formats 2, 2a or 2b allocated to the mobile is located by the parameter $n_{PUCCH}^{(2)}$, communicated to the mobile by the RRC *ConnectionReconfiguration* message.

The PUCCH physical channel in format 3 allocated to the mobile is located by the parameter $n_{\text{PUCCH}}^{(3)}$, the mobile was earlier configured with a list of 4 values $n_{\text{PUCCH}}^{(3)}$ by an RRC message.

The parameter $n_{\text{PUCCH}}^{(3)}$ is allocated to the mobile by the Transmit Power Control (TPC) command of the DCI information in format 3, transmitted in the PDCCH physical channel, the TPC command is in this case interpreted differently.

9.2.7. *Power control*

Average power P_{PUCCH}, expressed in dBm, of the PUCCH physical channel is provided by the following formula:

$$P_{\text{PUCCH}}(i) = \min\left\{P_{\text{CMAX}}(i), P_{\text{CALCULATED}}(i)\right\}$$

$P_{\text{CMAX}}(i)$ corresponds to the configured maximum power of the mobile, for the *i* subframe.

$$P_{\text{CALCULATED}}(i) = P_{\text{O\_PUCCH}} + PL + h\left(n_{CQI}, n_{HARQ,} n_{SR}\right) \\ +\Delta_{\text{F\_PUCCH}}(F) + \Delta_{TxD}(F') + g(i)$$

$P_{\text{O\_PUCCH}}$ corresponds to the power received by the eNB entity, for a resource block, for the PUCCH physical channel in formats 1, 1a and 1b.

$P_{\text{O\_PUCCH}}$ corresponds to the power amount $P_{\text{O\_NOMINAL\_PUCCH}}$ communicated by the SIB2 system information and $P_{\text{O\_UE\_PUCCH}}(j)$ by an RRC message.

PL corresponds to an assessment of the propagation loss calculated by the mobile.

PL is the difference between the value of the power of the reference signal communicated by the SIB2 system information or an RRC message on the one hand, and the value of the Reference Signal Received Power (RSRP) on the other hand.

$h(n_{CQI}, n_{HARQ}, n_{SR})$ introduces a power offset which depends on the number of bits transmitted in the PUCCH physical channel.

$\Delta_{\mathrm{F\_PUCCH}}(F)$ introduces a power offset which depends on the format of the PUCCH physical channel.

$\Delta_{TxD}(F')$ corresponds to a power offset introduced in Release 10, to take into consideration the transmit diversity.

$g(i)$ corresponds to the power control communicated by the eNB entity in the DCI information in formats 1, 1A, 1B, 1D, 2, 2A, 3 and 3A.

9.3. PUSCH physical channel

The Physical Uplink Shared Channel (PUSCH) transmits Uplink Shared Channel (UL-SCH) and Uplink Control Information (UCI).

The UCI control information is comprised of HARQ Indicator (HI), Channel Quality Indicator (CQI), Rank Indicator (RI) and Precoder Matrix Indicator (PMI).

For Release 8 and 9, simultaneous transmission of the PUSCH physical channel and Physical Uplink Control Channel (PUCCH) is not supported.

The UCI information transmission in the PUSCH physical channel is carried out, on the one hand, jointly with the traffic data transfer or Radio Resource Control (RRC) message, and, on the other hand, for the transfer of the aperiodic reports of UCI information.

For Release 10, the simultaneous transmission of the PUSCH and PUCCH physical channels is supported.

The UCI information transmission in the PUCCH physical channel is maintained when traffic data or RRC control must be transferred.

The processing associated with the PUSCH physical channel is summarized in Figure 9.10.

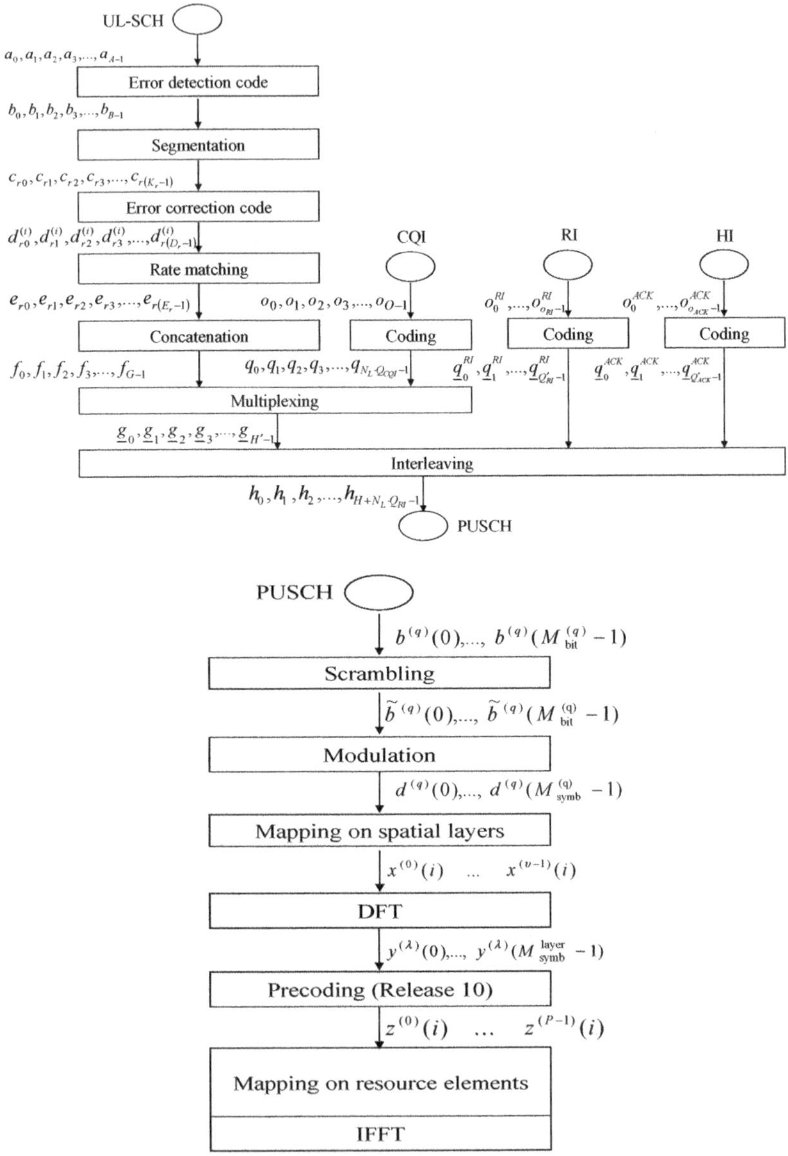

Figure 9.10. *Processing associated with the PUSCH physical channel*

The UL-SCH transport channel contains the traffic data (IP packets) and of RRC control corresponding to following messages:

– the common or dedicated control messages;

– the transport messages of the Non-Access Stratum (NAS) signaling.

For Release 8 and 9, the UL-SCH transport channel is in the form of a transport block.

For Release 10, the UL-SCH transport channel is in the form of one or two transport blocks.

9.3.1. *Error detection code*

The error detection is obtained from a Cyclic Redundancy Check (CRC) code.

The CRC structure is the remainder of the block transport division $a_0, a_1, a_2, a_3, ..., a_{A-1}$ by a generator polynomial, and the remainder $p_0, p_1, p_2, p_3, ..., p_{L-1}$ is made of the cyclic redundancy bits.

The concatenation of the transport block $a_0, a_1, a_2, a_3, ..., a_{A-1}$ and of the 24 bits of the cyclic redundancy $p_0, p_1, p_2, p_3, ..., p_{L-1}$ forms the input structure $b_0, b_1, b_2, b_3, ..., b_{B-1}$ of the segmentation.

9.3.2. *Segmentation*

The maximum size of the block processed by the error correction code is equal to 6144 bits.

If the block $b_0, b_1, b_2, b_3, ..., b_{B-1}$ is of a size larger than this value, it must be segmented and each segment must have its own error detection code whose length is equal to 24 bits for making the sequence $c_{r0}, c_{r1}, c_{r2}, c_{r3}, ..., c_{r(K_r-1)}$, with r and K_r respectively corresponding to the number of the segment and the number of the bits of the segment.

9.3.3. *Error correction code*

When the segmentation is carried out on the sequence $b_0, b_1, b_2, b_3, ..., b_{B-1}$, the error correction mechanism is applicable to each segment.

The error correction code is a turbo code made of a Parallel Concatenated Convolutional Code (PDCCC) and of a Quadratic Permutation Polynomial (QPP) interleaving (Figure 9.11).

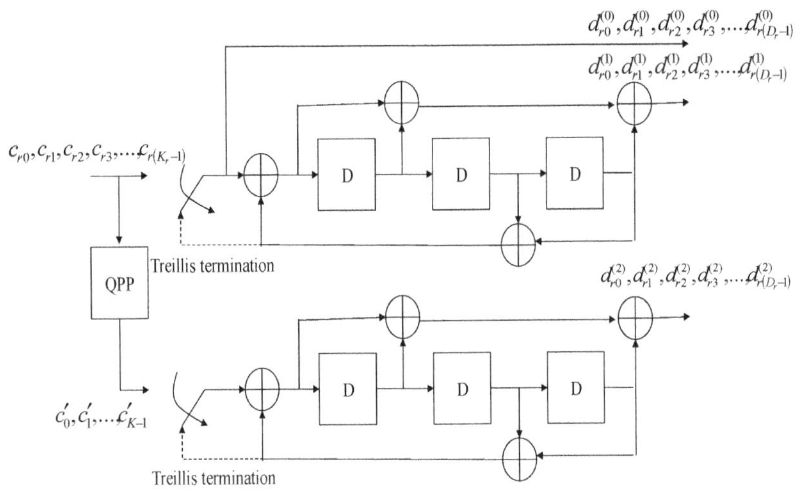

Figure 9.11. *Turbo code*

The error correction code produces three sequences $d_{r0}^{(i)}, d_{r1}^{(i)}, d_{r2}^{(i)}, d_{r3}^{(i)}, ..., d_{r(D_r-1)}^{(i)}$, with $i = 0,1$ and 2 and D_r corresponding to the number of sequence bits.

The sequence $d_{r0}^{(0)}, d_{r1}^{(0)}, d_{r2}^{(0)}, d_{r3}^{(0)}, ..., d_{r(D_r-1)}^{(0)}$ is equal to the sequence $c_{r0}, c_{r1}, c_{r2}, c_{r3}, ..., c_{r(K_r-1)}$.

The input of the first encoder is the data structure $c_{r0}, c_{r1}, c_{r2}, c_{r3}, ..., c_{r(K_r-1)}$. The sequence $d_{r0}^{(1)}, d_{r1}^{(1)}, d_{r2}^{(1)}, d_{r3}^{(1)}, ..., d_{r(D_r-1)}^{(1)}$ is produced from the first encoder.

The input of the second encoder is the output of the QPP interleaving $c'_0, c'_1, ..., c'_{K-1}$. The sequence $d_{r0}^{(2)}, d_{r1}^{(2)}, d_{r2}^{(2)}, d_{r3}^{(2)}, ..., d_{r(D_r-1)}^{(2)}$ is produced from the second encoder.

9.3.4. *Rate matching*

The propagation conditions determine the coding scheme retained. The rate matching determines the sequence with the desired coding rate.

The three sequences issued by the turbo code are interleaved then stored in a circular memory.

The circular memory bits are selected or used to form the output sequence $e_{r0}, e_{r1}, e_{r2}, e_{r3}, ..., e_{r(E_r-1)}$, with E_r corresponding to the number of the bits of the sequence.

The Hybrid Automatic Repeat reQuest (HARQ) mechanism is integrated with processing of rate matching.

The output sequence of the circular memory is transmitted from a starting position given by the Redundancy Version (RV).

A different redundancy version facilitates the retransmission of selected data.

For the Chase Combining mechanism, the retransmission contains the same sequence. The improvement of the error correction code is due to the increase of the signal-to-noise ratio.

For the incremental redundancy mechanism, the retransmission contains some different sequences. The improvement of the error correction code is obtained thanks to the increase of the number of redundancy bits.

9.3.5. *Concatenation*

The different sequences $e_{r0}, e_{r1}, e_{r2}, e_{r3}, ..., e_{r(E_r-1)}$, with $r = 0, ..., C-1$, C corresponding to the number of segments, are concatenated to form the sequence $f_0, f_1, f_2, f_3, ..., f_{G-1}$, with G corresponding to the number of sequence bits.

9.3.6. *Coding of the control data*

The coding of HI, RI and CQI/PMI control information is carried out independently.

9.3.6.1. *Coding of HI information*

The HI information involves the positive ACK or negative NACK acknowledgment of data received on the Physical Downlink Shared Channel (PDSCH).

The HI information is made of one bit $[o_0^{ACK}]$ or two bits $[o_0^{ACK}\ o_1^{ACK}]$ according to the number of transport blocks generated by the Downlink Shared Channel (DL-SCH).

The number of coded bits depends on the index of modulation Q_m used for the PUSCH physical channel:

– $Q_m = 2$ corresponds to Quadrature Phase-Shift Keying (QPSK) modulation;

– $Q_m = 4$ corresponds to the 16-Quadrature Amplitude Modulation (16-QAM);

– $Q_m = 6$ corresponds to the 64-QAM modulation.

Table 9.8 provides HI information coding, in the case where one bit $[o_0^{ACK}]$ is transmitted.

Q_m	Coding
2	$[o_0^{ACK}$ y]
4	$[o_0^{ACK}$ y x x]
6	$[o_0^{ACK}$ y x x x x]

Table 9.8. *HI information coding: a bit is transmitted*

Table 9.9 provides HI information coding, in the case where two bits $[o_0^{ACK}\ o_1^{ACK}]$ are transmitted.

Q_m	Coding
2	$[o_0^{ACK}\ o_1^{ACK}\ o_2^{ACK}\ o_0^{ACK}\ o_1^{ACK}\ o_2^{ACK}]$
4	$[o_0^{ACK}\ o_1^{ACK}$ x x $o_2^{ACK}\ o_0^{ACK}$ x x $o_1^{ACK}\ o_2^{ACK}$ x x]
6	$[o_0^{ACK}\ o_1^{ACK}$ x x x x $o_2^{ACK}\ o_0^{ACK}$ x x x x $o_1^{ACK}\ o_2^{ACK}$ x x x x]

$o_2^{ACK} = (o_0^{ACK} + o_1^{ACK}) \bmod 2$

Table 9.9. *HI information coding: two bits are transmitted*

x and y are spaces reserved for scrambling of the bit $[o_0^{ACK}]$ or of the bits $[o_0^{ACK}\ o_1^{ACK}]$ in order to maximize the Euclidian distance of the modulation symbols transporting HI information.

The sequence $q_0^{ACK}, q_1^{ACK}, q_2^{ACK}, \ldots, q_{Q_{ACK}-1}^{ACK}$ is a concatenation of the multiple blocks, Q_{ACK} representing the bits total number.

For Release 10, the sequence $o_0^{ACK}\ o_1^{ACK}, \ldots, o_{O^{ACK}-1}^{ACK}$ of the HI information, is made of 11 bits for the Frequency Division Duplex (FDD) mode or of 21 bits for the Time Division Duplex (TDD) mode.

If the number of bits is less than or equal to 11, the sequence $o_0^{ACK}\ o_1^{ACK}, \ldots, o_{O^{ACK}-1}^{ACK}$ of the HI information is coded to generate the sequence $\tilde{q}_0^{ACK}\ \tilde{q}_1^{ACK}, \ldots, \tilde{q}_{31}^{ACK}$ with $\tilde{q}_i^{ACK} = \sum_{n=0}^{O^{ACK}-1} \left(o_n^{ACK} \cdot M_{i,n}\right) \bmod 2$.

i takes the values between 0 and 31.

The sequence values $M_{i,n}$ are shown in Table 9.10.

The sequence $q_0^{ACK}, q_1^{ACK}, q_2^{ACK}, \ldots, q_{Q_{ACK}-1}^{ACK}$ is a circular repetition of the $\tilde{q}_0^{ACK}\ \tilde{q}_1^{ACK}, \ldots, \tilde{q}_{31}^{ACK}$ sequence.

I	$M_{i,0}$	$M_{i,1}$	$M_{i,2}$	$M_{i,3}$	$M_{i,4}$	$M_{i,5}$	$M_{i,6}$	$M_{i,7}$	$M_{i,8}$	$M_{i,9}$	$M_{i,10}$
0	1	1	0	0	0	0	0	0	0	0	1
1	1	1	1	0	0	0	0	0	0	1	1
2	1	0	0	1	0	0	1	0	1	1	1
3	1	0	1	1	0	0	0	0	1	0	1
4	1	1	1	1	0	0	0	1	0	0	1
5	1	1	0	0	1	0	1	1	1	0	1
6	1	0	1	0	1	0	1	0	1	1	1
7	1	0	0	1	1	0	0	1	1	0	1
8	1	1	0	1	1	0	0	1	0	1	1
9	1	0	1	1	1	0	1	0	0	1	1
10	1	0	1	0	0	1	1	1	0	1	1
11	1	1	1	0	0	1	1	0	1	0	1
12	1	0	0	1	0	1	0	1	1	1	1
13	1	1	0	1	0	1	0	1	0	1	1
14	1	0	0	0	1	1	0	1	0	0	1
15	1	1	0	0	1	1	1	1	0	1	1
16	1	1	1	0	1	1	1	0	0	1	0
17	1	0	0	1	1	1	0	0	1	0	0
18	1	1	0	1	1	1	1	1	0	0	0
19	1	0	0	0	0	1	1	0	0	0	0
20	1	0	1	0	0	0	1	0	0	0	1
21	1	1	0	1	0	0	0	0	0	1	1
22	1	0	0	0	1	0	0	1	1	0	1
23	1	1	1	0	1	0	0	0	1	1	1
24	1	1	1	1	1	0	1	1	1	1	0
25	1	1	0	0	0	1	1	1	0	0	1
26	1	0	1	1	0	1	0	0	1	1	0
27	1	1	1	1	0	1	0	1	1	1	0
28	1	0	1	0	1	1	1	0	1	0	0
29	1	0	1	1	1	1	1	1	1	0	0
30	1	1	1	1	1	1	1	1	1	1	1
31	1	0	0	0	0	0	0	0	0	0	0

Table 9.10. *Sequences* $M_{i,n}$

If the number of bits is greater than 11, the sequence is divided into two:

$$o_0^{ACK}, o_1^{ACK}, o_2^{ACK}, \ldots, o_{\lceil O^{ACK}/2 \rceil -1}^{ACK} \text{ and } o_{\lceil O^{ACK}/2 \rceil}^{ACK}, o_{\lceil O^{ACK}/2 \rceil +1}^{ACK}, o_{\lceil O^{ACK}/2 \rceil +2}^{ACK}, \ldots, o_{O^{ACK}-1}^{ACK}.$$

Each part of the sequence $o_0^{ACK}\ o_1^{ACK},...,o_{O^{ACK}-1}^{ACK}$ is coded to generate:

$$\tilde{q}_i = \sum_{n=0}^{\lceil O^{ACK}/2 \rceil - 1} \left(o_n^{ACK} \cdot M_{i,n} \right) \bmod 2 \text{ and } \tilde{\tilde{q}}_i = \sum_{n=0}^{O^{ACK} - \lceil O^{ACK}/2 \rceil - 1} \left(o_{\lceil O^{ACK}/2 \rceil + n}^{ACK} \cdot M_{i,n} \right) \bmod 2$$

The sequence $q_0^{ACK}, q_1^{ACK}, q_2^{ACK}, ..., q_{Q_{ACK}-1}^{ACK}$ is a concatenation of the $\tilde{q}_0^{ACK}\ \tilde{q}_1^{ACK},...,\tilde{q}_{31}^{ACK}$ and $\tilde{\tilde{q}}_0^{ACK}\ \tilde{\tilde{q}}_1^{ACK},...,\tilde{\tilde{q}}_{31}^{ACK}$ sequences.

Before carrying out the multiplexing, the $q_0^{ACK}, q_1^{ACK}, q_2^{ACK}, ..., q_{Q_{ACK}-1}^{ACK}$ sequence is vectorized to form the sequence vector $\underline{q}_0^{ACK}, \underline{q}_1^{ACK}, ..., \underline{q}_{Q'_{ACK}-1}^{ACK}$, $\underline{q}_i^{ACK}$ corresponding to a column vector and $Q'_{ACK} = Q_{ACK} / Q_m$.

9.3.6.2. *Coding of RI information*

The RI information determines the number of spatial layers recommended for the PDSCH physical channel.

For Release 8 and 9, the RI information is made of a bit $[o_0^{RI}]$ if the maximum number of spatial layers is equal to 2 or two bits $[o_0^{RI}\ o_1^{RI}]$ if the maximum number of spatial layers is equal to 4.

For Release 10, the RI information is made of three bits $[o_0^{RI}\ o_1^{RI}\ o_2^{RI}]$ if the maximum number of spatial layers is equal to 8.

Table 9.11 provides the RI information coding, in the case where one bit $[o_0^{RI}]$ is transmitted.

Q_m	Coding
2	$[o_0^{RI}$ y]
4	$[o_0^{RI}$ y x x]
6	$[o_0^{RI}$ y x x x x]

Table 9.11. *RI information coding: one bit is transmitted*

Table 9.12 provides the RI information coding, in the case where two bits $[o_0^{RI}\ o_1^{RI}]$ or three bits $[o_0^{RI}\ o_1^{RI} o_2^{RI}]$ are transmitted.

Q_m	Coding
2	$[o_0^{RI}\ o_1^{RI}\ o_2^{RI}\ o_0^{RI}\ o_1^{RI}\ o_2^{RI}]$
4	$[o_0^{RI}\ o_1^{RI}\ \mathrm{x\,x}\ o_2^{RI}\ o_0^{RI}\ \mathrm{x\,x}\ o_1^{RI}\ o_2^{RI}\ \mathrm{x\,x}]$
6	$[o_0^{RI}\ o_1^{RI}\ \mathrm{x\,x\,x\,x}\ o_2^{RI}\ o_0^{RI}\ \mathrm{x\,x\,x\,x}\ o_1^{RI}\ o_2^{RI}\ \mathrm{x\,x\,x\,x}]$

$o_2^{RI} = (o_0^{RI} + o_1^{RI}) \bmod 2$ for Release 8 and 9

Table 9.12. *RI information coding: two bits are transmitted*

The sequence $q_0^{RI}, q_1^{RI}, q_2^{RI}, \ldots, q_{Q_{RI}-1}^{RI}$ is a concatenation of multiple coding blocks, Q_{RI} representing the bits total number.

If the number of bits is less than or equal to 11, the sequence $[o_0^{RI}\ o_1^{RI}, \ldots, o_{O^{RI}-1}^{RI}]$ of the RI information is coded to generate the sequence $[\tilde{q}_0^{RI}\ \tilde{q}_1^{RI}, \ldots, \tilde{q}_{31}^{RI}]$, with $\tilde{q}_i^{RI} = \sum_{n=0}^{O^{RI}-1} \left(o_n^{RI} \cdot M_{i,n}\right) \bmod 2$.

i takes the values between 0 and 31.

The sequence values $M_{i,n}$ are shown in Table 9.10.

The sequence $q_0^{RI}, q_1^{RI}, q_2^{RI}, \ldots, q_{Q_{RI}-1}^{RI}$ is a circular repetition of the $\tilde{q}_0^{RI}\ \tilde{q}_1^{RI}, \ldots, \tilde{q}_{31}^{RI}$ sequence.

If the number of bits is greater than 11, the sequence is divided into two:

$$o_0^{RI}, o_1^{RI}, o_2^{RI}, \ldots, o_{\lceil O^{RI}/2 \rceil -1}^{RI} \text{ and } o_{\lceil O^{RI}/2 \rceil}^{RI}, o_{\lceil O^{RI}/2 \rceil +1}^{RI}, o_{\lceil O^{RI}/2 \rceil +2}^{RI}, \ldots, o_{O^{RI}-1}^{RI}.$$

Each part of the sequence $[o_0^{RI}\ o_1^{RI}, \ldots, o_{O^{RI}-1}^{RI}]$ is coded to generate:

$$\tilde{q}_i = \sum_{n=0}^{\lceil O^{RI}/2 \rceil -1} \left(o_n^{RI} \cdot M_{i,n}\right) \bmod 2 \text{ and } \tilde{\tilde{q}}_i = \sum_{n=0}^{O^{RI}-\lceil O^{RI}/2 \rceil -1} \left(o_{\lceil O^{RI}/2 \rceil +n}^{RI} \cdot M_{i,n}\right) \bmod 2$$

The sequence $q_0^{RI}, q_1^{RI}, q_2^{RI}, ..., q_{Q_{RI}-1}^{RI}$ is a concatenation of the $\tilde{q}_0^{RI}\ \tilde{q}_1^{RI}, ..., \tilde{q}_{31}^{RI}$ and $\tilde{\tilde{q}}_0^{RI}\ \tilde{\tilde{q}}_1^{RI}, ..., \tilde{\tilde{q}}_{31}^{RI}$ sequences.

Before carrying out the multiplexing, the sequence $q_0^{RI}, q_1^{RI}, q_2^{RI}, ..., q_{Q_{RI}-1}^{RI}$ is vectorized to form the sequence vector $\underline{q}_0^{RI}, \underline{q}_1^{RI}, ..., \underline{q}_{Q'_{RI}-1}^{RI}$, $\underline{q}_i^{RI}$ corresponding to a column vector and $Q'_{RI} = Q_{RI} / Q_m$

9.3.6.3. *Coding of CQI/PMI information*

The CQI information represents the Modulation and Coding Scheme (MCS) recommended for the PDSCH physical channel.

The PMI information provides indications related to the precoding matrix in the case of using spatial multiplexing in closed loop.

The sequence of the bits at the input of the encoder is written as $o_0, o_1, o_2, o_3, ..., o_{O-1}$, O corresponding to the number of sequence bits.

If the number of bits O is less than or equal to 11, the sequence $o_0, o_1, o_2, o_3, ..., o_{O-1}$ of the CQI information is coded to generate the sequence $[\tilde{q}_0\ \tilde{q}_1, ..., \tilde{q}_{31}]$ with $\tilde{q}_i = \sum_{n=0}^{O-1} (o_n \cdot M_{i,n}) \bmod 2$.

i takes the values between 0 and 31.

The sequence values $M_{i,n}$ are shown in Table 9.10.

If the number of bits is greater than 11, the coding involves the following operations:

– an error detection code obtained from a cyclic redundancy code;

– an error correction code obtained from a triple convolution code;

– a rate matching.

The coding sequence of the CQI/PMI information bits, written as $q_0, q_1, q_2, q_3, ..., q_{N_L \cdot Q_{CQI}-1}$, is a circular repetition of the CQI/PMI coded information blocks.

9.3.7. *Multiplexing of the control and traffic data*

The data sequence originating from CQI/PMI control $q_0, q_1, q_2, q_3, ..., q_{N_L \cdot Q_{CQI}-1}$ and the sequence originating from the UL-SCH transport channel $f_0, f_1, f_2, f_3, ..., f_{G-1}$ are multiplexed to produce the column vector sequence $\underline{g}_0, \underline{g}_1, \underline{g}_2, \underline{g}_3, ..., \underline{g}_{H'-1}$.

$$H = \left(G + N_L \cdot Q_{CQI}\right)$$

$$H' = H / \left(N_L \cdot Q_m\right)$$

If the UL-DCH physical channel is made of two transport blocks, the multiplexing is carried out with the transport block using the best modulation and coding scheme.

If the modulation and coding schemes are identical for the two blocks, the multiplexing is carried out with the first transport block.

9.3.8. *Interleaving*

The interleaving is carried out on the $\underline{q}_0^{ACK}, \underline{q}_1^{ACK}, ..., \underline{q}_{Q'_{ACK}-1}^{ACK}$, $\underline{q}_0^{RI}, \underline{q}_1^{RI}, ..., \underline{q}_{Q'_{RI}-1}^{RI}$ and $\underline{g}_0, \underline{g}_1, \underline{g}_2, \underline{g}_3, ..., \underline{g}_{H'-1}$ sequences to form sequence $h_0, h_1, h_2, ..., h_{H+Q_{RI}-1}$.

9.3.9. *Scrambling*

The sequence of bits $b^{(q)}(0), ..., b^{(q)}(M_{\text{bit}}^{(q)}-1)$, $M_{\text{bit}}^{(q)}$ designating the number of bits from the code word q, is scrambled by sequence $c^q(i)$.

For Release 8 and 9, the index of the code word q is equal to 0.

For Release 10, the index of the code word is of the value 0 or 1.

It results in the bit sequence $\tilde{b}^{(q)}(0), ..., \tilde{b}^{(q)}(M_{\text{bit}}^{(q)}-1)$, with $\tilde{b}^q(i) = \left(b^q(i) + c^q(i)\right) \bmod 2$, with the following restrictions:

– if $b^{(q)}(i) = \text{x}$, then $\tilde{b}^{(q)}(i) = 1$;

– if $b^{(q)}(i) = \text{y}$, then $\tilde{b}^{(q)}(i) = \tilde{b}^{(q)}(i-1)$.

The scrambling sequence is made of a pseudo-random sequence that uses a 31-bit polynomial.

The initial value of the scrambling sequence c_{init} is calculated from the following formula:

$$c_{\text{init}} = n_{\text{RNTI}} \cdot 2^{14} + q \cdot 2^{13} + \lfloor n_s/2 \rfloor \cdot 2^9 + N_{\text{ID}}^{\text{cell}}$$

n_{RNTI} corresponds to the Radio Network Temporary Identifier (RNTI) allocated to the mobile during connection.

n_s corresponds to the number of the time slot of the time frame.

$N_{\text{ID}}^{\text{cell}}$ corresponds to the Physical-layer Cell Identity (PCI).

9.3.10. *Modulation*

The propagation conditions determine the modulation scheme retained: QPSK, 16-QAM or 64-QAM.

The sequence of $d^{(q)}(0),...,d^{(q)}(M_{\text{symb}}^{(q)}-1)$ symbols, originating from the modulation, is generated from the bit sequence $\tilde{b}^{(q)}(0),...,\tilde{b}^{(q)}(M_{\text{bit}}^{(q)}-1)$.

9.3.11. *Mapping on the spatial layers*

The sequence of symbols $d^{(q)}(0),...,d^{(q)}(M_{\text{symb}}^{(q)}-1)$ is mapped in the sequence $x^{(0)}(i)$ ... $x^{(\upsilon-1)}(i)$, with $i = 0,1,...,M_{\text{symb}}^{\text{layer}}-1$, υ corresponding to the number of spatial layers and $M_{\text{symb}}^{\text{layer}}$ corresponding to the number of symbols per layer.

For Release 8 and 9, when a single spatial layer is used, the following relationship ensues: $x^{(0)}(i) = d^{(0)}(i)$.

For Release 10, in the case of using the spatial multiplexing, the symbols are mapped on one to four spatial layers (Figure 9.12 and 9.13).

In case of using the transmit diversity, a single code word is used and the symbols are mapped in two or four spatial layers.

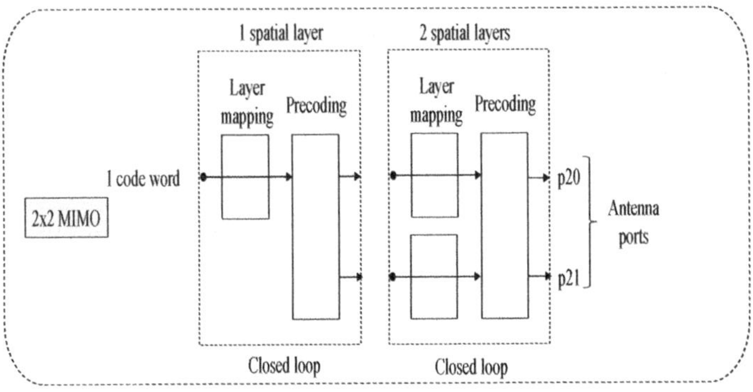

Figure 9.12. *Layer mapping: 2x2 MIMO*

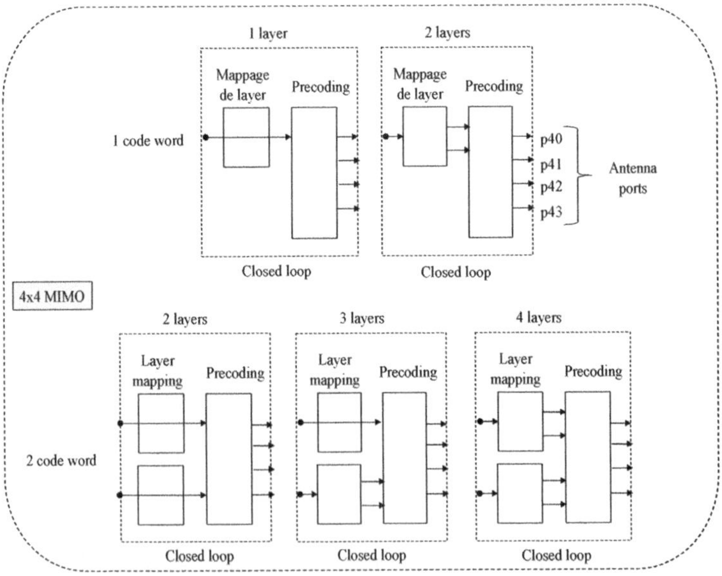

Figure 9.13. *Layer mapping: 4x4 MIMO*

9.3.12. *Precoding*

For each layer $\lambda = 0,1,...,\upsilon-1$, the sequence of symbols $x^{(\lambda)}(0),...,x^{(\lambda)}(M_{\text{symb}}^{\text{layer}}-1)$ is divided into $M_{\text{symb}}^{\text{layer}}/M_{\text{sc}}^{\text{PUSCH}}$ sets, $M_{\text{sc}}^{\text{PUSCH}}$ corresponding to the number of subcarriers of the set, each set corresponding to an Orthogonal Frequency Division Multiplexing (OFDM) symbol.

A Fast Fourier Transform (FFT) is applied to each set.

The size $M_{\text{sc}}^{\text{PUSCH}}$ must satisfy the following relationship:

$$M_{\text{sc}}^{\text{PUSCH}} = N_{\text{sc}}^{\text{RB}} \times 2^{\alpha_2} \times 3^{\alpha_3} \times 5^{\alpha_5} \leq N_{\text{sc}}^{\text{RB}} \times N_{\text{RB}}^{\text{UL}}.$$

$N_{\text{sc}}^{\text{RB}}$ corresponds to the number of subcarriers per resource block, in the frequency domain.

$N_{\text{RB}}^{\text{UL}}$ corresponds to the number of resource blocks allocated to the mobile.

$\alpha_2, \alpha_3, \alpha_5$ are integers.

There is a sequence of symbols $y^{(\lambda)}(0),...,y^{(\lambda)}(M_{\text{symb}}^{\text{layer}}-1)$ obtained from the following formula:

$$y^{(\lambda)}(l \cdot M_{\text{sc}}^{\text{PUSCH}} + k) = \frac{1}{\sqrt{M_{\text{sc}}^{\text{PUSCH}}}} \sum_{i=0}^{M_{\text{sc}}^{\text{PUSCH}}-1} x^{(\lambda)}(l \cdot M_{\text{sc}}^{\text{PUSCH}} + i) e^{-j\frac{2\pi ik}{M_{\text{sc}}^{\text{PUSCH}}}}$$

$k = 0,...,M_{\text{sc}}^{\text{PUSCH}}-1$, k corresponding to the index of the subcarrier in the frequency domain.

$l = 0,...,M_{\text{symb}}^{\text{layer}}/M_{\text{sc}}^{\text{PUSCH}}-1$, l corresponding to the index of the OFDM symbol.

For Release 10, an additional precoding is applied to the sequence $y^{(0)}(i)$ ... $y^{(\upsilon-1)}(i)$, in the case of a spatial multiplexing, to generate the symbol sequence $z^{(0)}(i)$ ... $z^{(P-1)}(i)$, P corresponding to the number of antenna port.

9.3.13. *Mapping on the resource elements*

The mapping of the PUSCH physical channel on the Resource Elements (RE) is described in Figure 9.14.

The HI information is localized at the end of OFDM symbols, on both sides of the Demodulation Reference Signal (DM-RS).

The RI information is localized at the end of OFDM symbols, on both sides of the HI information.

The CQI/PMI information is localized at the beginning of the OFDM symbols and uses the set of available OFDM symbols.

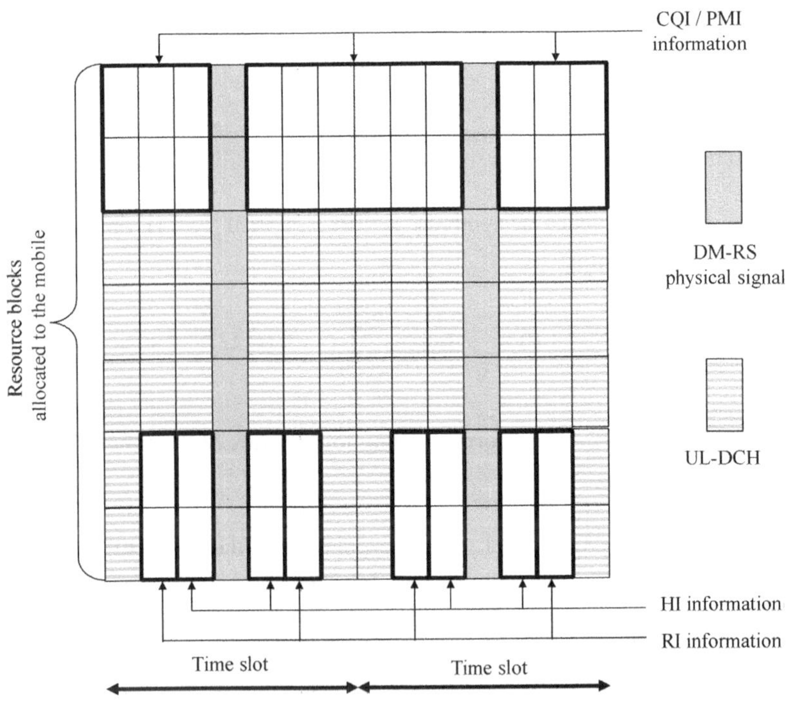

Figure 9.14. *Mapping of the PUSCH physical signal*

9.3.14. *Allocation of resources*

9.3.14.1. *Type-0 allocation of resources*

The type-0 allocation of resources is indicated in the Physical Downlink Control Channel (PDCCH), by the Downlink Control Information (DCI) in format 0 and 4.

The type-0 allocation of resources allocates a set of Virtual Resource Blocks (VRB), mapped on the Physical Resource Blocks (PRB).

When allocating resources, the Resource Indication Value (RIV) parameter is used. It corresponds to a Resource Block (RB) from the start RB_{start} and to a number L_{CRBs} of adjacent virtual resource blocks.

$$RIV = N_{\mathrm{RB}}^{\mathrm{UL}}(L_{\mathrm{CRBs}} - 1) + RB_{\mathrm{START}}, \text{ if } (L_{\mathrm{CRBs}} - 1) \leq \left\lfloor N_{\mathrm{RB}}^{\mathrm{UL}} / 2 \right\rfloor$$

Otherwise, $RIV = N_{\mathrm{RB}}^{\mathrm{UL}}(N_{\mathrm{RB}}^{\mathrm{UL}} - L_{\mathrm{CRBs}} + 1) + (N_{\mathrm{RB}}^{\mathrm{UL}} - 1 - RB_{\mathrm{START}})$

$N_{\mathrm{RB}}^{\mathrm{UL}}$ corresponds to the bandwidth of the radio channel, expressed in the number of resource block allocated.

9.3.14.2. *Type-1 allocation of resources*

The type-1 allocation of resources is indicated in the PDCCH physical channel by the DCI information in formats 0 and 4.

The type-1 allocation of resources, introduced in Release 10, facilitates the allocation of two sets of adjacent resource blocks, each set comprises one or several Resource Blocks Groups (RBG).

The size (P) of the resource block group depends on the bandwidth of the radio channel (Table 9.13).

The type-1 allocation of resources is defined by 4 indices, s1, s2, s3 and s4:

– s_0 and $s_1 - 1$ respectively define the first and the last resource block group of the first set;

– s_2 and $s_3 - 1$ respectively define the first and the last resource block group of the second set.

The four indices are combined in an index r, defined as a sum of binomial coefficients:

$$r = \sum_{i=0}^{M-1} \left\langle \begin{matrix} N - s_i \\ M - i \end{matrix} \right\rangle, \text{ with } M=4 \text{ and } N = \left\lceil N_{RB}^{UL} / P \right\rceil + 1 \text{ (Table 9.13).}$$

Radio channel	1.4-MHz	3-MHz	5-MHz	10-MHz	15-MHz	20-MHz
N_{RB}^{UL}	6	15	25	50	75	100
P	1	2	2	3	4	4
N	7	9	14	18	20	26

Table 9.13. *Characteristics of the type-1 allocation of resources*

Figure 9.15 describes an example of the type-1 allocation of resources for a 3-MHz bandwidth radio channel.

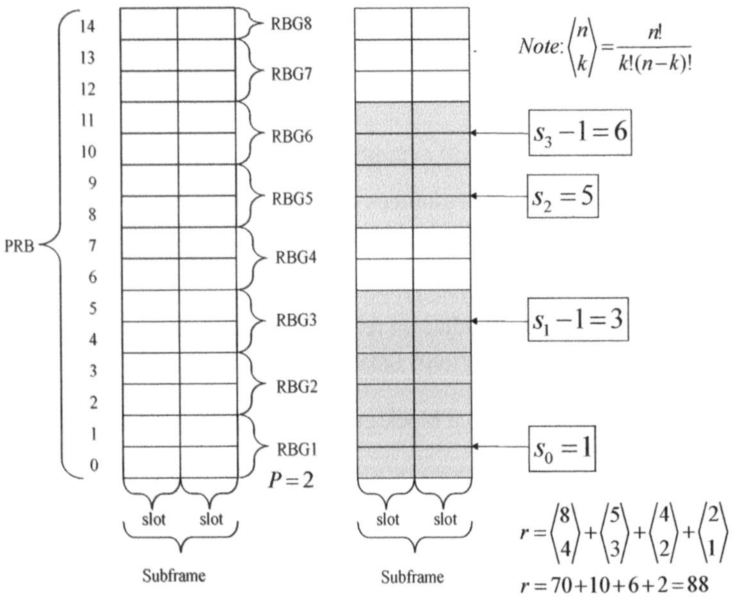

Figure 9.15. *Type-1 allocation of resources*

9.3.15. *Frequency hopping*

The frequency hopping is activated if the *Frequency hopping flag* bit of the DCI information in format 0 is set at ONE and only for the type-0 allocation of resources.

According to the radio channel bandwidth, the field *Resource allocation* of the DCI information in format 0 contains 1 or 2 bits whose value determines the type of frequency hopping (Table 9.14).

The type-1 frequency hopping is a hopping through fixed frequency offset.

The type-2 frequency hopping is a frequency hopping through predefined pattern.

Radio channel	Number of bits	Value	Type of frequency hopping
1.4-MHz 3, 5-MHz	1	0	Type 1
		1	Type 2
10, 15, 20-MHz	2	00	Type 1
		01	
		10	
		11	Type 2

Table 9.14. *Bits of the Resource allocation field*

9.3.15.1. *Type-1 frequency hopping*

The type-1 frequency hopping is carried out between the frequency subbands of the radio channel, whose number depends on the bandwidth of the radio channel (Table 9.15).

The frequency hopping is intra-subframe when the allocation defines the position of the resource blocks for the first time slot of the subframe, the frequency hopping is carried out for the second time slot.

The frequency hopping is inter-subframe when the allocation defines the position of the resource blocks for the even-numbered transmissions; the frequency hopping is carried out for the odd-numbered transmissions.

The frequency hopping intra- and inter-subframe provides the diversity in the case of a non-adaptive retransmission of the same transport block.

The maximum number of adjacent resource blocks is limited to $\lfloor 2^y / N_{\mathrm{RB}}^{\mathrm{UL}} \rfloor$, with $y = \lceil \log_2(N_{\mathrm{RB}}^{\mathrm{UL}}(N_{\mathrm{RB}}^{\mathrm{UL}}+1)/2) \rceil - N_{UL\_hop}$.

y corresponds to the size of the *Resource allocation* field, excluding the one or two bits of frequency hopping information.

$N_{\mathrm{RB}}^{\mathrm{UL}}$ corresponds to the bandwidth of the radio channel, expressed in the number of resource blocks.

N_{UL_hop} corresponds to the number of bits (1 or 2) of frequency hopping information.

Bandwidth (MHz)	1.4	3	5	10	15	20
Number of subbands	2	2	2	4	4	4
Number of adjacent RB	2	4	10	10	13	20

Table 9.15. *Characteristics of the type-1 frequency hopping*

The position of the resource blocks which experience a frequency hopping, located by the index $\tilde{n}_{PRB}$, depends on the value of the bit or of the two bits N_{UL_hop} (Table 9.16).

Bandwidth	N_{UL_hop}	Value N_{UL_hop}	$\tilde{n}_{PRB}$
1.4-MHz 3, 5-MHz	1	0	$\left(\lfloor N_{RB}^{PUSCH}/2 \rfloor + \tilde{n}_{PRB}^{S1}\right) \bmod N_{RB}^{PUSCH}$
		1	Type 2
10, 15, 20-MHz	2	00	$\left(\lfloor N_{RB}^{PUSCH}/4 \rfloor + \tilde{n}_{PRB}^{S1}\right) \bmod N_{RB}^{PUSCH}$
		01	$\left(-\lfloor N_{RB}^{PUSCH}/4 \rfloor + \tilde{n}_{PRB}^{S1}\right) \bmod N_{RB}^{PUSCH}$
		10	$\left(\lfloor N_{RB}^{PUSCH}/2 \rfloor + \tilde{n}_{PRB}^{S1}\right) \bmod N_{RB}^{PUSCH}$
		11	Type 2

Table 9.16. *Index* $\tilde{n}_{PRB}$

N_{RB}^{PUSCH} corresponds to the number of resource blocks allocated to the PUSCH physical channel.

$\tilde{n}_{PRB}^{S1}$ corresponds to the index of the position, in the PUSCH physical channel, of the resource block that does not experience the frequency hopping.

Figure 9.16 describes the type-1 frequency hopping for a 3-MHz bandwidth.

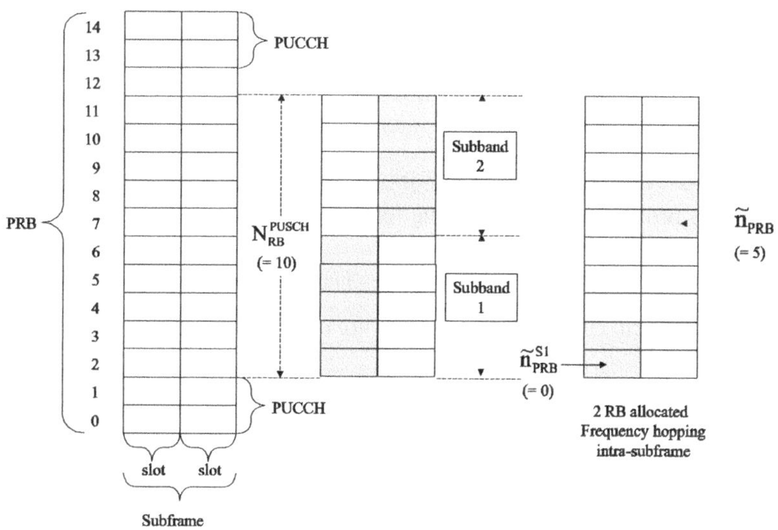

Figure 9.16. *Type-1 frequency hopping for a 3-MHz bandwidth*

9.3.15.2. *Type-2 frequency hopping*

The type-2 frequency hopping is carried out between the frequency sub-bands of the radio channel, whose number can be 1 to 4 in value and is configured by the RRC *ConnexionReconfiguration* message.

The maximum number of resource blocks allocated to a mobile depends on the number of subbands and on the number of resource blocks allocated to the PUSCH physical channel.

The frequency hopping can be implemented by the subband hopping or by the mirror effect when a single subband is configured.

The subband hopping consists of changing the subband for each time slot according to a predefined pattern, the position of the resource blocks allocated to the mobile remains fixed in connection with the subband.

The mirror effect consists of positioning the resource blocks allocated to the mobile symmetrically in relation to the center of the radio channel.

The type-2 frequency hopping is available in intra-subframe, in inter-subframe and for an intra-subframe and inter-subframe combination.

9.3.16. *Power control*

Average power P_{PUSCH}, expressed in dBm, of the PUSCH physical channel is provided by the following formula:

$P_{\text{PUSCH}}(i) = \min\{P_{\text{CMAX}}(i), P_{\text{CALCULATED}}(i)\}$, in case the PUSCH and PUCCH physical channel transmissions are not simultaneous.

$P_{\text{PUSCH}}(i) = 10\log_{10}\left(\hat{P}_{\text{CMAX}}(i) - \hat{P}_{\text{PUCCH}}(i)\right)$, in case the transmissions of PUSCH and PUCCH physical channels are simultaneous (Release 10).

$$P_{\text{CALCULATED}}(i) = 10\log_{10}(M_{\text{PUSCH}}(i)) + P_{\text{O\_PUSCH}}(j)$$

$$+\alpha(j)\cdot PL + \Delta_{\text{TF}}(i) + f(i)$$

$P_{\text{CMAX}}(i)$ corresponds to the configured maximum power of the mobile, for the i subframe.

$\hat{P}_{\text{CMAX}}(i)$ corresponds to the maximum power expressed in linear unit.

$\hat{P}_{\text{PUCCH}}(i)$ corresponds to the power of the PUCCH physical channel, expressed in linear unit.

$M_{\text{PUSCH}}(i)$ corresponds to the bandwidth allocated to the mobile, expressed in the number of resource blocks allocated.

$P_{O\_PUSCH}(j)$ corresponds to the power received by the evolved Node B (eNB) entity, for a resource block.

The index (j) corresponds to the type of scheduling: semi-persistent, dynamic or random access.

$\alpha(j)$ corresponds to a configuration parameter of the fractional power control.

PL corresponds to an assessment of the propagation loss calculated by the mobile.

PL is the difference between the value of the power of the reference signal communicated by the SIB2 system information or a RRC message, on one hand, and the value of the Reference Signal Received Power (RSRP), on the other hand.

$\Delta_{TF}(i)$ is a parameter that depends on the modulation and coding scheme. The value of this parameter increases with the modulation index.

$f(i)$ corresponds to the Transmit Power Control (TPC) communicated by the eNB entity in the DCI information in formats 0, 3, 3A and 4.

$f(i)$ either represents an absolute command of power control, or an accumulation of the commands received.

10

Radio Interface Procedures

10.1. Access control

10.1.1. *PRACH physical channel acquisition*

The procedure of access to the evolved Node B (eNB) entity is developed for accessing Physical Random Access Channel (PRACH) that the mobile is going to use to carry out the random access.

The different physical channels and physical signals processed by the mobile for the acquisition of the PRACH physical channel can be found in Table 10.1.

Physical channel/signal	Acquisition of the parameters
PSS	Frequency synchronization Time synchronization (half-frame) Parameter $N_{ID}^{(2)}$
SSS	Time synchronization (frame) Length of the cyclic prefix Parameter $N_{ID}^{(1)}$
Cell-Specific RS	Coherent demodulation
PBCH	Bandwidth of the channel for the downlink direction Parameter of the PHICH physical channel
PCFICH	Size of the PDCCH physical channel
PDCCH	Detection of the SI-RNTI identifier
PDSCH	SIB1 and SIB2 system information

Table 10.1. *Acquisition of the PRACH physical channel*

The Primary Synchronization Signal (PSS) facilitates the following functions:

– frequency synchronization;

– time synchronization at the level of the Orthogonal Frequency Division Multiplexing (OFDM) symbol, of the time slot (periodicity 0.5 ms), of the subframe (periodicity 1 ms) and the half-frame (periodicity 5 ms);

– determination of the value of the number $N_{ID}^{(2)}$.

The Secondary Synchronization Signal (SSS) facilitates the following functions:

– time synchronization at the frame level (periodicity 10 ms);

– determination of the Frequency Division Duplex (FDD) or Time Division Duplex (TDD) mode;

– determination of the type of the Cyclic Prefix (CP), normal or extended;

– determination of the value of the number $N_{ID}^{(1)}$.

The number of the Physical-layer Cell Identity (PCI) is equal to $N_{ID}^{cell} = 3N_{ID}^{(1)} + N_{ID}^{(2)}$.

$N_{ID}^{(1)}$ represents the group number and can take the values between 0 and 167. It is determined by the SSS physical signal.

$N_{ID}^{(2)}$ represents the number in the group and can be a value between 0 and 2. It is determined by the PSS physical signal.

The value N_{ID}^{cell} determines the mapping of the Cell-Specific Reference Signal (RS) on the Resource Elements (RE).

The Cell-Specific RS physical signal is used to carry out the coherent signal demodulation.

After having withdrawn the resource elements allocated to the Cell-Specific RS physical signal, the mobile can analyze the Physical Broadcast Channel (PBCH) which transmits the Broadcast Channel (BCH) containing the system information corresponding to the Master Information Block (MIB) message.

The MIB message provides the information which allows the mobile to subsequently analyze the Physical Control Format Indicator Channel (PCFICH) and the Physical Downlink Control Channel (PDCCH):

– the bandwidth of the radio channel, for the downlink direction;

– the System Frame Number (SFN);

– the configuration of the Physical HARQ Indication Channel (PHICH).

The processing of the PSS and SSS physical signals on the one hand, and of the PBCH physical channel on the other hand, is independent of the bandwidth of the radio channel.

After having withdrawn the resource elements allocated to the Cell-Specific RS physical signal, the mobile can analyze the PCFICH physical channel which, for each subframe, defines the number of OFDM symbols allocated to the PDCCH physical channel.

After having withdrawn the resource elements allocated to the Cell-Specific RS physical signal and to the PCFICH and PHICH physical channels, the mobile can analyze the PDCCH physical channel.

The PDCCH physical channel transports the information which facilitates the recovery of the System Information Block 1 and 2 (SIB1 and SIB2) messages contained in the Physical Downlink Shared Channel (PDSCH), from the detection of the System Information Radio Network Temporary Identifier (SI-RNTI).

The SIB1 message provides the following information:

– the bandwidth of the radio channel, for the uplink direction;

– the configuration of the type-2 time frame;

– the scheduling of the other SIB system information.

The SIB2 message provides the information related to the configuration of radio resources allocated to the PRACH physical channel.

10.1.2. *Random access*

The random access procedure is initialized by the mobile and is required by the following cases:

– when establishing the connection to the eNB entity;

– when changing the cell during the session (handover);

– when updating Timing Advance (TA);

– when re-establishing the connection to the eNB entity.

The random access procedure is said to be with contention when the mobile chooses the used resources (PRACH channel, preamble) randomly. This occurs for the establishment or re-establishment phases of the connection.

The random access procedure is said to be without contention when the eNB entity is providing the resource to be used to the mobile. This happens for the handover or for the updating of the timing advance.

10.1.2.1. *Random access with contention*

The random access procedure with contention, during the establishment or re-establishment of the eNB entity connection, is described in Figure 10.1.

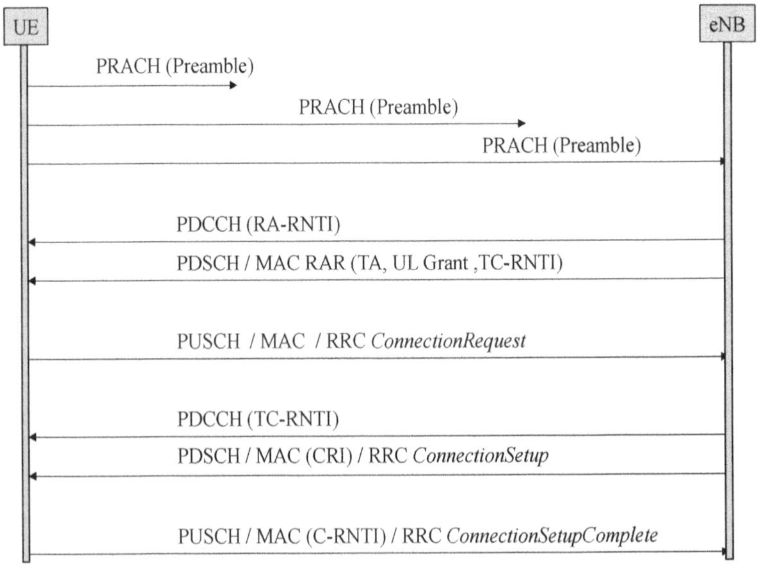

Figure 10.1. *Random access with contention*

The mobile transmits the preamble in the PRACH physical channel, this preamble is chosen randomly in a list communicated by the eNB entity in the SIB2 system information.

In case the eNB entity does not respond, the mobile retransmits the preamble while increasing the transmission power. The maximum number of retransmission is shown by the SIB2 system information or the RRC *ConnectionReconfiguration* message.

The risk of contention is linked to the fact that several mobiles can access the same PRACH physical channel and use the same preamble.

When the eNB entity receives the PRACH physical channel, it calculates the timing advance and transmits to the mobile:

– the Downlink Control Information (DCI) in the PDCCH physical channel, in format 1A or 1C:

- the mobile recovers DCI information from RA-RNTI identity,

- the mobile recovers the description of its data in the PDSCH physical channel;

– the Random Access Response (MAC RAR) frame containing the index of the preamble, the timing advance, the resource to use (UL Grant) for the transmission in the Physical Uplink Shared Channel (PUSCH) and the Temporary Cell RNTI (TC-RNTI):

- this identity is temporary because several mobiles may consider that this identity is allocated to them, thereby leading to contention.

The RA-RNTI identity is calculated in the following way:

$$RA\text{-}RNTI = 1 + t\_id + 10*f\_id$$

t_id is the index of the first subframe used by the PRACH physical channel ($0 \leq t\_id < 10$).

f_id is the index of the PRACH physical channel, in the frequency domain ($0 \leq f\_id < 6$).

The mobile initializes its timing advance and responds with the RRC *ConnectionRequest* message containing:

– the Shortened Temporary Mobile Subscriber Identity (S-TMSI), if the mobile is already connected;

– a random number in the contrary case.

When the eNB entity receives the RRC *ConnectionRequest* message, it transmits to the mobile:

– the DCI information in the PDCCH physical channel:

- the mobile recovers DCI information from TC-RNTI identity,

- the mobile recovers the description of its data in the PDSCH physical channel;

– the heading MAC RAR containing the UE CRI (Contention Resolution Identity) control element:

- this control element reproduces the identity of the RRC *ConnectionRequest* message, thereby resolving the contention issue;

– the RRC *ConnectionSetup* message.

The TC-RNTI temporary identity becomes the C-RNTI definitive identity allocated to mobile.

The mobile shows its C-RNTI identifier in the control element of the MAC frame and confirms the connection through the RRC *ConnectionSetupComplete* message.

10.1.2.2. *Random access without contention*

The random access procedure without contention, when changing the cell during the session, is described in Figure 10.2.

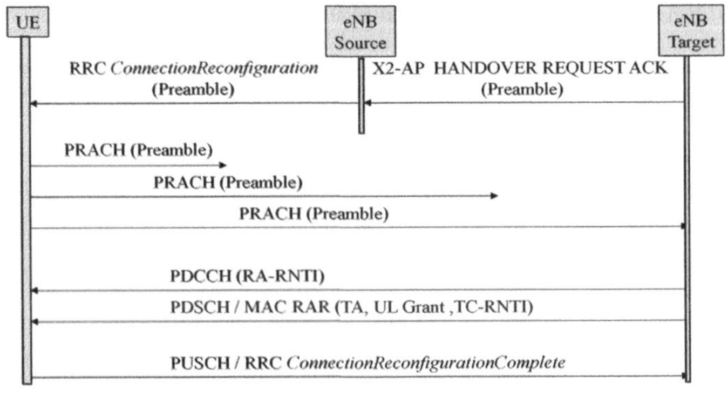

Figure 10.2. *Random access without contention, in case of handover*

During the procedure of handover based on the interface X2, the target eNB entity provides the characteristics of the radio interface to the source eNB entity in a message X2-AP HANDOVER REQUEST ACK.

This message contains the information element *Handover Command* which specifies the preamble that the mobile must use during the random access procedure to the eNB entity.

The source eNB entity transfers the information element *Handover Command* in the RRC *ConnectionReconfiguration* message which triggers the handover for the mobile.

The random access procedure as before is comprised of the preamble transmission by the mobile and the MAC RAR frame by the eNB entity.

The mobile connection is finalized when the mobile transmits the RRC *ConnectionReconfigurationComplete* message.

The random access procedure without contention, during the updating of the timing advance, is described in Figure 10.3.

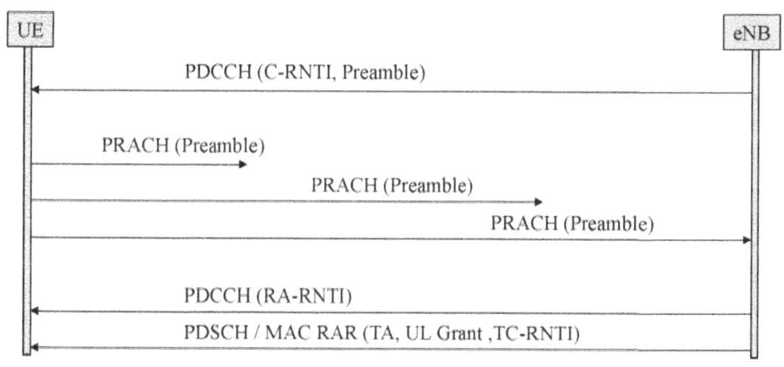

Figure 10.3. *Random access without contention, in case of updating the timing advance*

The eNB entity transmits the DCI information in the PDCCH physical channel in format 1A, to the mobile.

The mobile recovers DCI information from C-RNTI identity.

The mobile recovers the resource that it must use for the transmission of the preamble in the PRACH physical channel.

The random access procedure as before is comprised of the preamble transmission by the mobile and of the MAC RAR frame by the eNB entity.

When the eNB entity receives the acknowledgement of the MAC RAR frame, it can proceed to the transmission of the data for the uplink or the downlink direction.

10.2. Data transfer

10.2.1. *Scheduling*

The data scheduling is the operation carried out by the evolved Node B (eNB) entity which consists of providing Resource Blocks (RB) to the mobiles and of controlling transmission power, for the downlink and uplink direction.

In the time domain, the allocation of the resources corresponds to a subframe of a Transmission Time Interval (TTI) duration of a millisecond, which represents two resource blocks.

In the frequency domain, the eNB entity can give the mobile several resource blocks, each block corresponds to a frequency band of 180 kHz, formed from an Orthogonal Frequency Division Multiplexing (OFDM) of 12 subcarriers with 15 kHz spacing or 24 subcarriers with 7.5 kHz spacing.

In the space domain, the mobile can receive and transmit different resource blocks, simultaneously and in the same frequency band, thanks to the Multiple Input Multiple Output (MIMO) mechanism:

– 2×2 MIMO or 4×4 MIMO for the downlink direction in Release 8;

– 8×8 MIMO for the downlink direction and 2×2 MIMO or 4×4 MIMO for the uplink direction in Release 10.

For the downlink direction, the scheduling algorithm takes into account the following information:

– the information recovered by the mobiles, in relation to the Channel Quality Indicator (CQI), to the Precoding Matrix Indicator (PMI) and to the number of spatial layers, from the signal received on the Physical Downlink Shared Channel (PDSCH);

– the information recovered by the adjacent eNB entities under the Inter Cell Interference Coordination (ICIC) mechanism, in relation to the Relative Narrowband Tx Power (RNTP);

– the information transmitted by the Mobility Management Entity (MME) in relation to the category of the mobile and to the QoS Class Identifier (QCI) level;

– the local information related to the state of the memory, to the need of retransmission, to the state of available resources and to the measure intervals constructed for the mobile.

For the uplink direction, the scheduling algorithm takes into account the following information:

– the information recovered by the mobiles, in relation to the Power Headroom Report (PHR) and the Buffer Status Report (BSR);

– the information recovered by the adjacent eNB entities under the ICIC mechanism, in relation to the Interference Overload Indication (IOI) and to High Interference Indication (HII);

– the information recovered by the MME entity in relation to the category of the mobile and to the QCI level;

– the local information related to the level of quality measured on the Sounding Reference Signal (SRS), to the need for retransmission, to the state of available resources and to the measure intervals constructed for the mobile.

The results of the scheduling make up the Downlink Control Information (DCI) communicated in the Physical Downlink Control Channel (PDCCH).

The DCI control information in formats 0 and 4 facilitates the scheduling of the data in the Physical Uplink Shared Channel (PUSCH).

The DCI control information in formats 1, 1A, 1B, 1C, 1D and 2, 2B, 2C facilitates the scheduling of data in the PDSCH physical channel.

The DCI control information in format 3 facilitates the Transmit Power Control (TPC) of the PUSCH physical channel and the Physical Uplink Control Channel (PUCCH).

The allocation of the resources is shown by the Radio Network Temporary Identifier (RNTI) specific to a mobile (C-RNTI, SPS C-RNTI, TPC-RNTI) or to a set of mobiles (P-RNTI, RA-RNTI, TC-RNTI, SI-RNTI).

In the Frequency Division Duplex (FDD) mode, the resource allocation for the uplink direction, signaled in the PDCCH physical channel of the subframe n, is applicable to the subframe n+4 ms.

In the Time Division Duplex (TDD) mode, the shift between signaling in the PDCCH physical channel and transmission in the PUSCH physical channel depends on the configuration of the time frame and on the number of the subframe of the PDCCH physical channel (Table 10.2).

Configuration of the time frame	Number of the subframe of the PDCCH physical channel									
	0	1	2	3	4	5	6	7	8	9
1	–	6	–	–	4	–	6	–	–	4
2	–	–	–	4	–	–	–	–	4	–
3	4	–	–	–	–	–	–	–	4	4
4	–	–	–	–	–	–	–	–	4	4
5	–	–	–	–	–	–	–	–	4	
6	7	7	–	–	–	7	7	–	–	5

Table 10.2. *Shift between signaling and transmission*

For configuration 0 of the time frame, the shift between signaling and transmission depends on the value of the two bits *Uplink Index* of DCI control information in formats 0 and 4:

– the most significant bit at ONE and the least significant bit at ZERO, the shift is provided in Table 10.3;

– the most significant bit at ZERO and the least significant bit at ONE, the shift is fixed at 7 ms;

– the most significant bit at ZERO and the least significant bit at ONE, the mobile can transmit in the two subframes previously defined.

Configuration of the time frame	Number of the subframe of the PDCCH physical channel									
	0	1	2	3	4	5	6	7	8	9
0	4	6	–	–	–	4	6	–	–	–

Table 10.3. *Shift for configuration 0 of the time frame*

10.2.2. *DRX function*

The Discontinuous Reception (DRX) function determines the moments when the mobile must analyze the PDCCH physical channel, which allows it to avoid processing this channel every millisecond and in this way to preserve the consumption of its battery (Figure 10.4).

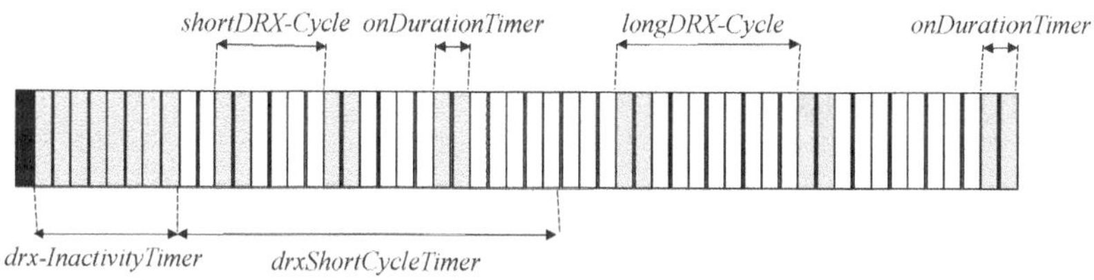

Figure 10.4. *DRX function*

An inactivity timer *drx-InactivityTimer* of the DRX function is triggered when the mobile receives data on the PDCCH physical channel.

This timer is reinitialized each time that the mobile receives data on the PDCCH physical channel.

When the timer *drx-InactivityTimer* expires, the mobile starts an optional period of short cycle corresponding to the timer *drxShortCycleTimer*.

During the short cycle, the mobile analyzes the PDCCH physical channel in a duration corresponding to the timer *onDurationTimer*.

During the short cycle, the triggering of the active period is provided by the following formula:

[(SFN * 10) + (the subframe number)] modulo (*shortDRX-Cycle*) = (*drxStartOffset*) modulo (*shortDRX-Cycle*)

When the timer *drxShortCycle Timer* expires, the mobile starts the long cycle period for which the triggering of the active period is provided by the following formula:

[(SFN * 10) + (subframe number)] modulo (*longDRX-Cycle*) = (*drxStartOffset*)

The parameter configuration of the DRX function is indicated in the Radio Resource Control (RRC) messages of establishment or re-establishment of the connection.

The eNB entity indicates the activation of the DRX function while using the DRX control element in the Medium Access Control (MAC) frame.

The *drx-RetransmissionTimer* timer defines an inactive period of the DRX function during an expected Hybrid Automatic Repeat reQuest (HARQ) retransmission.

The *drx-RetransmissionTimer* timer starts after the round trip HARQ timer associated with the HARQ retransmission.

10.2.3. *SPS function*

The Semi-Persistent Scheduling (SPS) function is applied to the applications that have a periodic character, for example the voice which produces a block every 20 ms.

The SPS function allows the eNB entity to avoid announcing the DCI control information in the PDCCH physical channel.

The SPS function is configured by the eNB entity in RRC messages of establishment or re-establishment of the connection or establishment of Data Radio Bearer (DRB) containing the following parameter:

– SPS Cell Radio Network Temporary Identifier (SPS C-RNTI);

– allocation periodicity, for example a transmission every 20 subframes for the voice;

– number of the HARQ processes of retransmission in case of errors, for the downlink direction.

When the SPS function has been configured, the eNB entity uses the DCI control information transmitted in the PDCCH physical channel to activate or release it.

When the SPS function has been configured, the mobile must nevertheless continue the analysis of the PDCCH physical channel, for the defined subframes by the DRX function, in order to detect the DCI control information in relation to the allocation release.

When the SPS function has been activated for the downlink direction, the subframe allocation of the PDSCH physical channel corresponds to the following formula:

$$(10*SFN+subframe)=[(10*SFN_{starttime}+subframe_{starttime}) +N*semiPersistSchedIntervalDL] \text{ modulo } 10240$$

The $SFN_{start\ time}$ and $subframe_{start\ time}$ parameters correspond to the values of the frame and subframe number when the SPS function has been activated.

The *semiPersistSchedIntervalDL* parameter corresponds to the allocation periodicity of resources for the downlink direction.

When the SPS function has been activated for the uplink direction, the subframe allocation of the PUSCH physical channel corresponds to the following formula:

$$(10*SFN+subframe)=[(10*SFN_{starttime}+subframe_{starttime}) +N*semiPersistSchedIntervalUL+Subframe\_Offset * (N \text{ modulo } 2)] \text{ modulo } 10240$$

The *semiPersistSchedIntervalUL* parameter corresponds to the allocation periodicity of resources for the uplink direction.

The Subframe_Offset parameter is optional and its value is equal to ZERO for the FDD mode and is indicated in Table 10.4 for the TDD mode.

Configuration of the subframe	Number of the subframe during the SPS activation	Subframe_Offset
0	Not applicable	0
1	2 and 7	1
	3 and 8	-1
2	2	5
	7	-5
3	2 and 3	1
	4	-2
4	2	1
	3	-1
5	Not applicable	0
6	Not applicable	0

Table 10.4. *Value of optional parameters Subframe_Offset*

10.2.4. *HARQ function*

In case of error, the retransmission implements two mechanisms:

– the Automatic Repeat reQuest (ARQ) mechanism elaborated by the RLC layer;

– the Hybrid ARQ (HARQ) mechanism established at the physical layer level, under the MAC layer control.

The ARQ mechanism takes over from the HARQ mechanism if the retransmissions at the physical layer level have failed.

The ARQ mechanism is only applied to the flow of traffic and signaling using the Acknowledged Mode (AM) of the Radio Link Control (RLC) protocol.

On the other hand, taking its speed into consideration, the HARQ mechanism is applied to the flow using the AM mode and Unacknowledged Mode (UM) of the RLC protocol.

Different redundancy version facilitates the retransmission of selected data:

– for the Chase Combining mechanism, the retransmission contains the same scheduling. The improvement of the error corrector code is due to the increase of the signal-to-noise ratio;

– for the incremental redundancy mechanism, the retransmission contains some different sequences. The improvement of the error corrector code is due to the increase of the number of redundancy bits.

The first transmission corresponding to the Redundancy Version 0 (RV0) contains the initial data to transmit and the redundancy bits generated by the error corrector code, whose number is determined by the coding rate:

$$\text{Coding rate} = N_{seq} / (N_{seq} + N_{red})$$

N_{seq} corresponds to the number of bits of the initial data;

N_{red} corresponds to the number of redundancy bits.

The HARQ mechanism functions with a window of one data unit. When a transport block is transmitted, the transmitter must wait for the reception of the acknowledgement before sending the following transport block.

This disposition penalizes the rate of the mobile and this restriction is removed thanks to the implementation of several HARQ parallel processes.

The combination of HARQ parallel processes and of the retransmission mechanism provokes a de-sequencing of the blocks received by the destination. The RLC layer assures the re-sequencing of different blocks received.

For the downlink direction, the HARQ mechanism is adaptive because the retransmission can modify the transmission parameters.

For the downlink direction, the HARQ mechanism is asynchronous because a HARQ process transmission is not constrained by an imposed period.

For the uplink direction, the HARQ mechanism is adaptive or non-adaptive. It is non-adaptive if the retransmission must use the same initial transmission parameters.

For the uplink direction, the HARQ mechanism is synchronous because the HARQ process transmission must be done according to a defined timing.

10.2.4.1. *Data transfer for the uplink direction*

The number of HARQ entities is double for the transmission mode with MIMO (mode 2) that of transmission mode without MIMO (mode 1), because one HARQ entity in the uplink transmission is allocated to each transport block:

– one transport block is used for transmission mode 1;

– two transport blocks are used for transmission mode 2.

In case of the Carrier Aggregation (CA), an HARQ entity is developed for each radio channel.

For the FDD mode, by the HARQ entity, the HARQ process number has a fixed value:

– value equal to 8 for transmission mode 1;

– value equal to 16 for transmission mode 2.

Figure 10.5 describes the HARQ mechanism for the transmission mode 1 in the FDD mode.

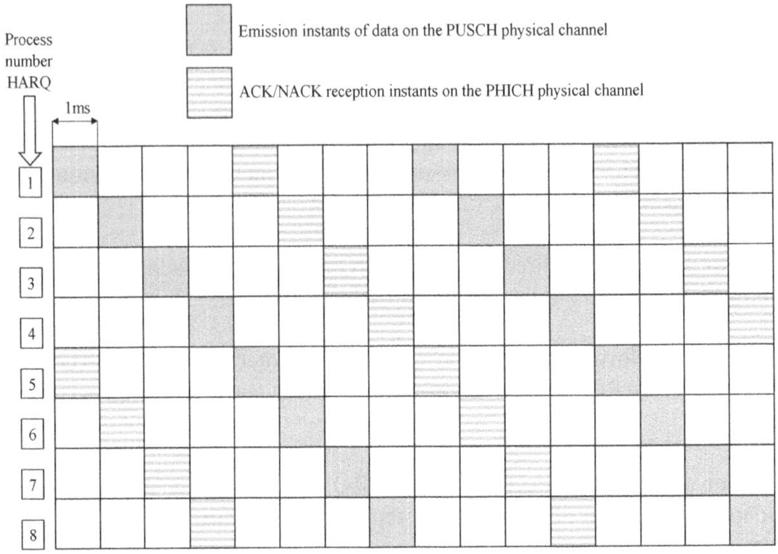

Figure 10.5. *HARQ function in the FDD mode: data transfer for the uplink direction*

The transmission moments for each HARQ process is synchronous, with an 8 ms periodicity.

The Physical HARQ Indicator Channel (PHICH) transports the HI indicator (ACK or NACK bits) for each transport block, 4 ms after the transmission of the data in the PUSCH physical channel.

Under a HARQ process, if a transport block is acknowledged, a new transport block can be transmitted 8 ms later. If a transport block is not acknowledged, the RV1 redundancy version is transmitted 8 ms later.

The HARQ processes used depend on the transmission moment of transport blocks. The 8 HARQ processes are systematically used by the mobile if the eNB entity allocates every millisecond of resources to the mobile.

For the TDD mode, the HARQ process number, under the HARQ entity depends on the type of configuration of the time frame and of the transmission mode (Table 10.5).

Configuration of the time frame	0	1	2	3	4	5	6
Number of subframes/uplink	6	4	2	3	2	1	5
Transmission 1 mode	7	4	2	3	2	1	6
Transmission 2 mode	14	8	4	3	4	2	12

Table 10.5. *HARQ process number in the TDD mode: data transfer for the uplink direction*

The shift which is given in milliseconds, between the transmission of a block in the PUSCH physical channel and the reception of the HI indicator in the PHICH physical channel is shown in Table 10.6.

Configuration of the time frame	Number of the subframe having received HI									
	0	1	2	3	4	5	6	7	8	9
0	7/6	4	–	–	–	7/6	4	–	–	–
1	–	4	–	–	6	–	4	–	–	6
2	–	–	–	6	–	–	–	–	6	–
3	6	–	–	–	–	–	–	–	6	6
4	–	–	–	–	–	–	–	–	6	6
5	–	–	–	–	–	–	–	–	6	–
6	6	4	–	–	–	7	4	–	–	6

Table 10.6. *Shift between the PUSCH and PHICH physical channels*

Figure 10.6 describes the HARQ mechanism for the configuration 1 of the time frame in the TDD mode.

The transport block transmitted in subframe 2 or 7 (subframe 3 or 8, respectively) receives the HI indicator with a shift of 4 ms (6 ms, respectively).

Under a HARQ process, if a transport block is acknowledged, a new transport block can be transmitted 10 ms later. If a transport block is not acknowledged, the RV1 redundancy version is transmitted 10 ms later.

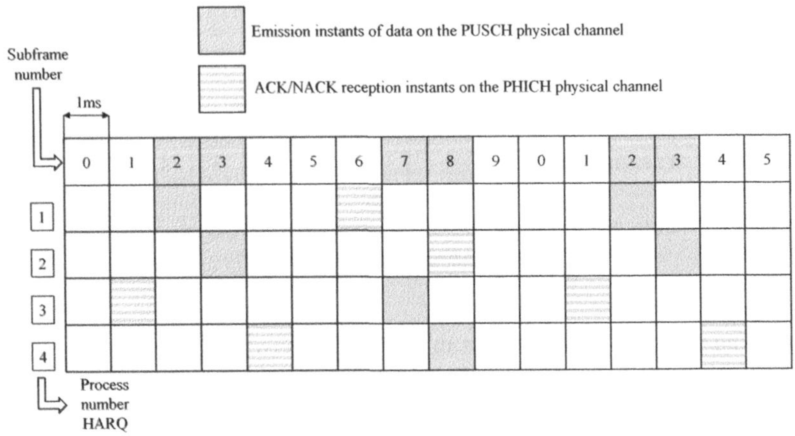

Figure 10.6. *HARQ function in the TDD mode, for configuration 1: data transfer for the uplink direction*

10.2.4.2. *Data transfer for the downlink direction*

Contrary to the data transfer for the uplink direction, a single HARQ entity is developed independently from the transmission mode.

In case of the aggregation of radio channels, a HARQ entity is developed for each radio channel.

Since the process number is shown in the DCI control information, the number of open HARQ processes is not fixed and depends on transmission needs.

For the FDD mode, the number of open HARQ processes is less than or equal to 8, the eight HARQ processes are only open if the eNB entity allocates resources to the mobile every millisecond.

The PUCCH or PUSCH physical channel transports the HI indicator with a shift of 4 ms after the transmission of the transport block in the PDSCH physical channel.

Under a HARQ process, if a transport block is acknowledged, a new transport block can be transmitted. If a transport block is not acknowledged, the RV1 redundancy version is transmitted. In the two cases, the moment used by the HARQ process is not imposed by a timing as for the uplink direction.

For the TDD mode, the HARQ process number, by the HARQ entity, depends on the configuration of the time frame (Table 10.7).

Configuration of the time frame	0	1	2	3	4	5	6
Number of subframes downlink direction	4	6	8	7	8	9	5
HARQ process number	≤ 4	≤ 7	≤ 10	≤ 9	≤ 12	≤ 15	≤ 6

Table 10.7. *HARQ process number in the TDD mode: data transfer for the downlink direction*

Table 10.8 shows the ACK / NACK number of bits to be transmitted, through the HARQ entity in the PUCCH physical channel, depending on the configuration of the time frame, for transmission mode 1. The indicated values are doubled in the case of transmission mode 2.

Configuration of the time frame	0	1	2	3	4	5	6
Number of subframes for the PUCCH physical channel	6	4	2	3	2	1	5
Number of ACK/NACK bits	4	6	8	7	8	9	5
Number of bits through subframe	1 or 0	1 or 2	4	2 or 3	4	9	1

Table 10.8. *Number of ACK/NACK bits to be transmitted in the PUCCH physical channel for the TDD mode and the transmission mode 1*

In certain cases, the capacity of the PUCCH physical channel is insufficient for recovering all the ACK/NACK information:

– format 1a has a capacity of one bit;

– format 1b has a capacity of 2 or 4 bits.

Two methods are defined in order to reduce the number of bits:

– coupling carried out from a logical AND of various ACK/NACK information:

- the NACK information of a transport block involves the retransmission of all the transport blocks which are linked to it,

- in case an ACK/NACK bit is generated by a subframe, the coupling generates a bit transmitted in format 1a,

- in case two ACK/NACK bits are generated by the subframe, the coupling generates two bits transmitted in format 1b (Figure 10.7);

– multiplexing of ACK/NACK information in the PUCCH physical channel:

- format 1 b is used for transporting 4 bits of information, each bit corresponding to ACK/NACK information generated by a subframe,

- in case two bits are generated by the subframe, the two ACK/NACK bits are linked by a logical AND (Figure 10.8),

- the NACK information of a transport block of a subframe involves the retransmission of the other transport block of the same subframe.

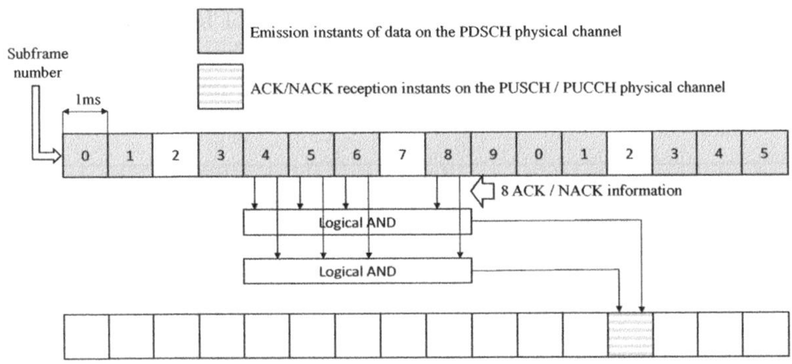

Figure 10.7. *Coupling ACK/NACK information: configuration 2 of the time frame Data transfer for the downlink direction*

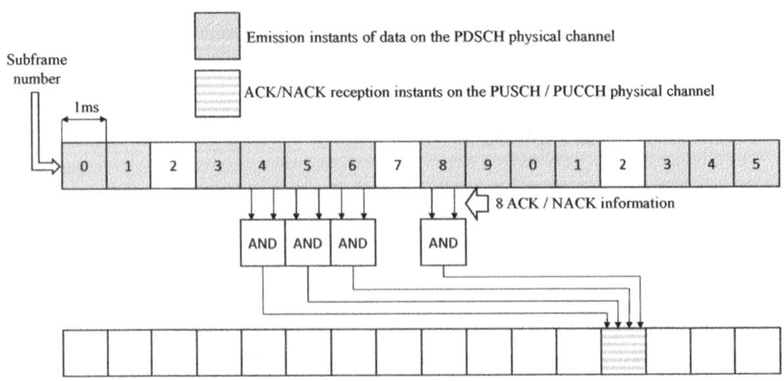

Figure 10.8. *Multiplexing ACK/NACK information: configuration 2 of the time frame Data transfer for the downlink direction*

10.2.5. *TTI bundling function*

Real-time applications like the voice are sensitive to packet jitter. The RTP protocol facilitates the correction of the jitter introduced in the network, with a maximum value of 40 ms and any packet having a jitter greater than this value is deleted.

The retransmission mechanism HARQ function of the HARQ function results in an increase of the packet jitter for each retransmission (for example, 8 ms in the FDD mode). The HARQ mechanism may be counter-productive in this case.

The Transmission Time Interval (TTI) bundling function consists of transmitting, in four consecutive subframes, four redundancy versions, without waiting for the return of HI information, in order to reduce the value of the jitter.

The same modulation and coding scheme and the same frequency band are used for transmission of the four redundancy versions.

The TTI bundling function is limited to the transmission for the uplink direction.

The TTI bundling function is not supported for the aggregation of radio channels.

In the FDD mode, the number of HARQ processes is reduced to 4 (Figure 10.9). The HI information is transmitted 4 ms after the last redundancy version. If the HI information corresponds to a NACK, the four redundancy versions are retransmitted with a 16 ms delay.

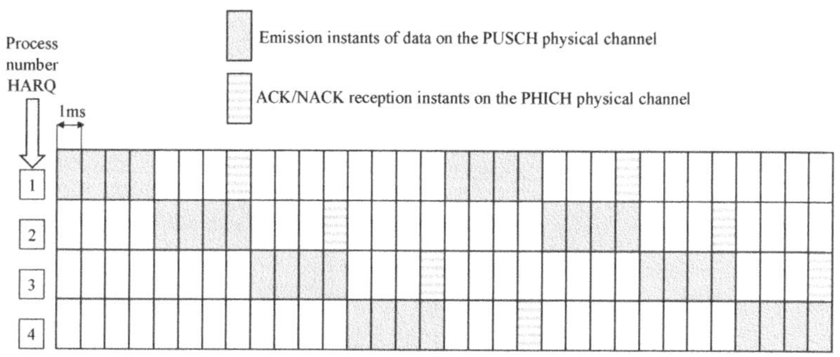

Figure 10.9. *TTI bundling function in the FDD mode*

In the TDD mode, the TTI bundling function is only applicable in configurations 0, 1 and 6 of the time frame.

The HARQ processes number is limited to 3 for the configurations 0 and 6 and to 2 for the configuration 1 (Figure 10.10).

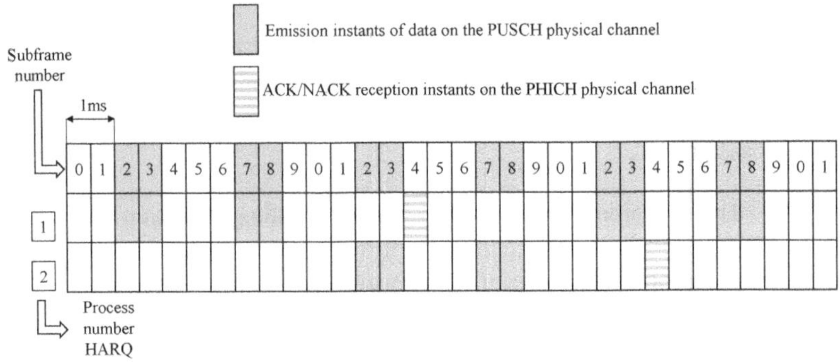

Figure 10.10. *TTI function bundling in the TDD mode for configuration 1*

Bibliography

Specifications

The general characteristics of the EPS network are described in the following specification:

3GPP TS 23.401
Technical Specification Group Services and System Aspect;
General Packet Radio Service (GPRS) enhancements for Evolved Universal Terrestrial Radio Access Network (E-UTRAN) access (Release 10)

The general characteristics of the MBMS network are described in the following specification:

3GPP TS 23.246
Technical Specification Group Services and System Aspects;
Multimedia Broadcast/Multicast Service (MBMS);
Architecture and functional description (Release 10)

The general characteristics of the LCS network are described in the following specification:

3GPP TS 23.271
Technical Specification Group Services and System Aspects;
Functional stage 2 description of Location Services (LCS) (Release 10)

The general characteristics of the radio interface are described in the following specifications:

3GPP TS 36.300
Technical Specification Group Radio Access Network;

Evolved Universal Terrestrial Radio Access (E-UTRA) and Evolved Universal Terrestrial Radio Access Network (E-UTRAN); Overall description; Stage 2 (Release 10)

3GPP TS 36.306
Technical Specification Group Radio Access Network;
Evolved Universal Terrestrial Radio Access (E-UTRA);
User Equipment (UE) radio access capabilities (Release 10)

The general characteristics of the NAS protocol are described in the following specification:

3GPP TS 24.301
Technical Specification Group Core Network and Terminals;
Non-Access-Stratum (NAS) protocol for Evolved Packet System (EPS);
Stage 3 (Release 10)

The general characteristics of the RRC protocol are described in the following specification:

3GPP TS 36.331
Technical Specification Group Radio Access Network;
Evolved Universal Terrestrial Radio Access (E-UTRA)
Radio Resource Control (RRC); Protocol specification (Release 10)

The characteristics of the data-link layer are described in the following specifications:

3GPP TS 36.323
Technical Specification Group Radio Access Network;
Evolved Universal Terrestrial Radio Access (E-UTRA);
Packet Data Convergence Protocol (PDCP) specification (Release 10)

3GPP TS 36.322
Technical Specification Group Radio Access Network;
Evolved Universal Terrestrial Radio Access (E-UTRA);
Radio Link Control (RLC) protocol specification (Release 10)

3GPP TS 36.321
Technical Specification Group Radio Access Network;
Evolved Universal Terrestrial Radio Access (E-UTRA);
Medium Access Control (MAC) protocol specification (Release 10)

The characteristics of the physical layer, the physical channels and the physical signals are described in the following specifications:

3GPP TS 36.101
Technical Specification Group Radio Access Network;
Evolved Universal Terrestrial Radio Access (E-UTRA);
User Equipment (UE) radio transmission and reception (Release 10)

3GPP TS 36.211
Technical Specification Group Radio Access Network;
Evolved Universal Terrestrial Radio Access (E-UTRA);
Physical Channels and Modulation (Release 10)

3GPP TS 36.212
Technical Specification Group Radio Access Network;
Evolved Universal Terrestrial Radio Access (E-UTRA);
Multiplexing and channel coding (Release 10)

3GPP TS 36.213
Technical Specification Group Radio Access Network;
Evolved Universal Terrestrial Radio Access (E-UTRA);
Physical layer procedures (Release 10)

Books

Readers can find additional information in the following publications that certainly cover the same themes, but with a different presentation.

JOHNSON C., *Long Term Evolution in Bullets*, Create Space Independent Publishing Platform, 2012.

BOUGUEN Y., HARDOUIN E., WOLFF F.X., *LTE and 4G Networks*, Eyrolles, September 2012.

DAHLMAN E., PARKVALL S., SKÖLD J., *4G LTE/LTE-Advanced for Mobile Broadband*, Elsevier, 2011.

SESIA S., TOUFIK I., BAKER M., *LTE – The UMTS Long Term Evolution: From Theory to Practice*, Wiley, 2009.

COX C., *An Introduction to LTE: LTE, LTE-Advanced, SAE and 4G Mobile Communications*, Wiley, 2012.

RUMNEY M., *LTE and the Evolution to 4G Wireless: Design and Measurement Challenges*, Wiley, 2013.

Tutorials

The following tutorials provide a summary of the different characteristics defined in *Releases* 8, 9 and 10.

3G Americas – The Mobile Broadband Evolution: 3GPP Release 8 and Beyond – HSPA1, SAE/LTE and LTE-Advanced – February 2009.

4G Americas – MIMO and Smart Antennas for Mobile Broadband Networks – June 2013.

4G Americas – 4G Mobile Broadband Evolution: 3GPP Release 10 and Beyond – HSPA1, SAE/LTE and LTE-Advanced – February 2011.

Rohde & Schwarz – Application Note – UMTS Long Term Evolution (LTE) – Technology Introduction – Gessner C., Roessler A., Kottkamp M. – July 2012.

Rohde & Schwarz – White Paper – LTE-Advanced Technology Introduction Meik Kottkamp – July 2010.

Keysight Technologies – Application Note – LTE-Advanced: Technology and Test Challenges – 3GPP Releases 10, 11, 12 and Beyond.

Agilent Technologies – Application Note – Introducing LTE-Advanced.

Index

D

E

F, G, H

I

K, L

M

N, O

P

T

U, V, X, Z

Other titles from

ISTE

in

Networks and Telecommunications

2015

BENSLAMA Malek, KIAMOUCHE Wassila, BATATIA Hadj
Connections Management Strategies in Satellite Cellular Networks

BENSLAMA Malek, BATATIA Hadj, BOUCENNA Mohamed Lamine
Ad Hoc Networks Telecommunications and Game Theory

BERTHOU Pascal, BAUDOIN Cédric, GAYRAUD Thierry, GINESTE Matthieu
Satellite and Terrestrial Hybrid Networks

LE RUYET Didier, PISCHELLA Mylène
Digital Communications 1: Source and Channel Coding

PISCHELLA Mylène, LE RUYET Didier
Digital Communications 2: Digital Modulations

PUJOLLE Guy
Software Networks

2014

ANJUM Bushra, PERROS Harry
Bandwidth Allocation for Video under Quality of Service Constraints

BATTU Daniel
New Telecom Networks: Enterprises and Security

BEN MAHMOUD Mohamed Slim, GUERBER Christophe, LARRIEU Nicolas, PIROVANO Alain, RADZIK José
Aeronautical Air–Ground Data Link Communications

BITAM Salim, MELLOUK Abdelhamid
Bio-inspired Routing Protocols for Vehicular Ad-Hoc Networks

CAMPISTA Miguel Elias Mitre, RUBINSTEIN Marcelo Gonçalves
Advanced Routing Protocols for Wireless Networks

CHETTO Maryline
Real-time Systems Scheduling 1: Fundamentals
Real-time Systems Scheduling 2: Focuses

EXPOSITO Ernesto, DIOP Codé
Smart SOA Platforms in Cloud Computing Architectures

MELLOUK Abdelhamid, CUADRA-SANCHEZ Antonio
Quality of Experience Engineering for Customer Added Value Services

OTEAFY Sharief M.A., HASSANEIN Hossam S.
Dynamic Wireless Sensor Networks

PEREZ André
Network Security

PERRET Etienne
Radio Frequency Identification and Sensors: From RFID to Chipless RFID

REMY Jean-Gabriel, LETAMENDIA Charlotte
LTE Standards
LTE Services

TANWIR Savera, PERROS Harry
VBR Video Traffic Models

VAN METER Rodney
Quantum Networking

XIONG Kaiqi
Resource Optimization and Security for Cloud Services

2013

ASSING Dominique, CALÉ Stéphane
Mobile Access Safety: Beyond BYOD

BEN MAHMOUD Mohamed Slim, LARRIEU Nicolas, PIROVANO Alain
Risk Propagation Assessment for Network Security: Application to Airport Communication Network Design

BEYLOT André-Luc, LABIOD Houda
Vehicular Networks: Models and Algorithms

BRITO Gabriel M., VELLOSO Pedro Braconnot, MORAES Igor M.
Information-Centric Networks: A New Paradigm for the Internet

BERTIN Emmanuel, CRESPI Noël
Architecture and Governance for Communication Services

DEUFF Dominique, COSQUER Mathilde
User-Centered Agile Method

DUARTE Otto Carlos, PUJOLLE Guy
Virtual Networks: Pluralistic Approach for the Next Generation of Internet

FOWLER Scott A., MELLOUK Abdelhamid, YAMADA Naomi
LTE-Advanced DRX Mechanism for Power Saving

JOBERT Sébastien *et al.*
Synchronous Ethernet and IEEE 1588 in Telecoms: Next Generation Synchronization Networks

MELLOUK Abdelhamid, HOCEINI Said, TRAN Hai Anh
Quality-of-Experience for Multimedia: Application to Content Delivery Network Architecture

NAIT-SIDI-MOH Ahmed, BAKHOUYA Mohamed, GABER Jaafar, WACK Maxime
Geopositioning and Mobility

PEREZ André
Voice over LTE: EPS and IMS Networks

2012

AL AGHA Khaldoun
Network Coding

BOUCHET Olivier
Wireless Optical Communications

DECREUSEFOND Laurent, MOYAL Pascal
Stochastic Modeling and Analysis of Telecoms Networks

DUFOUR Jean-Yves
Intelligent Video Surveillance Systems

EXPOSITO Ernesto
Advanced Transport Protocols: Designing the Next Generation

JUMIRA Oswald, ZEADALLY Sherali
Energy Efficiency in Wireless Networks

KRIEF Francine
Green Networking

PEREZ André
Mobile Networks Architecture

2011

BONALD Thomas, FEUILLET Mathieu
Network Performance Analysis

CARBOU Romain, DIAZ Michel, EXPOSITO Ernesto, ROMAN Rodrigo
Digital Home Networking

CHABANNE Hervé, URIEN Pascal, SUSINI Jean-Ferdinand
RFID and the Internet of Things

GARDUNO David, DIAZ Michel
Communicating Systems with UML 2: Modeling and Analysis of Network Protocols

LAHEURTE Jean-Marc
Compact Antennas for Wireless Communications and Terminals: Theory and Design

RÉMY Jean-Gabriel, LETAMENDIA Charlotte
Home Area Networks and IPTV

PALICOT Jacques
Radio Engineering: From Software Radio to Cognitive Radio

PEREZ André
IP, Ethernet and MPLS Networks: Resource and Fault Management

TOUTAIN Laurent, MINABURO Ana
Local Networks and the Internet: From Protocols to Interconnection

2010

CHAOUCHI Hakima
The Internet of Things

FRIKHA Mounir
Ad Hoc Networks: Routing, QoS and Optimization

KRIEF Francine
Communicating Embedded Systems / Network Applications

2009

CHAOUCHI Hakima, MAKNAVICIUS Maryline
Wireless and Mobile Network Security

VIVIER Emmanuelle
Radio Resources Management in WiMAX

2008

CHADUC Jean-Marc, POGOREL Gérard
The Radio Spectrum

GAÏTI Dominique
Autonomic Networks

LABIOD Houda
Wireless Ad Hoc and Sensor Networks

LECOY Pierre
Fiber-optic Communications

MELLOUK Abdelhamid
End-to-End Quality of Service Engineering in Next Generation Heterogeneous Networks

PAGANI Pascal *et al.*
Ultra-wideband Radio Propagation Channel

2007

BENSLIMANE Abderrahim
Multimedia Multicast on the Internet

PUJOLLE Guy
Management, Control and Evolution of IP Networks

SANCHEZ Javier, THIOUNE Mamadou
UMTS

VIVIER Guillaume
Reconfigurable Mobile Radio Systems

Printed and bound by CPI Group (UK) Ltd, Croydon, CR0 4YY

13/09/2026

14961017-0006